Nineteenth-Century Br

Histories of the Sacred and the Secular, 1700–2000
General Editor: **Professor David Nash**, Oxford Brookes University, UK

This series reflects the awakened and expanding profile of the history of religion within the academy in recent years. It intends publishing exciting new and high quality work on the history of religion and belief since 1700 and will encourage the production of interdisciplinary proposals and the use of innovative methodologies. The series will also welcome book proposals on the history of Atheism, Secularism, Humanism and unbelief/secularity and to encourage research agendas in this area alongside those in religious belief. The series will be happy to reflect the work of new scholars entering the field as well as the work of established scholars. The series welcomes proposals covering subjects in Britain, Europe, the United States and Oceana.

Titles include:

John Wolffe (*editor*)
IRISH RELIGIOUS CONFLICT IN COMPARATIVE PERSPECTIVE
Catholics, Protestants and Muslims

Clive D. Field
BRITAIN'S LAST RELIGIOUS REVIVAL?
Quantifying Belonging, Behaving and Believing in the Long 1950s

Jane Platt
THE ANGLICAN PARISH MAGAZINE 1859–1929

Jack Fairey
THE GREAT POWERS AND OTTOMAN CHRISTENDEM
The War for the Eastern Church in the Era of the Crimean War

Michael Rectenwald
NINETEENTH-CENTURY BRITISH SECULARISM
Science, Religion, and Literature

Histories of the Sacred and the Secular 1700–2000
Series Standing Order ISBN 978–1–137–32800–7 (hardback)
(*outside North America only*)

You can receive future titles in this series as they are published by placing a standing order. Please contact your bookseller or, in case of difficulty, write to us at the address below with your name and address, the title of the series and the ISBN quoted above.

Customer Services Department, Macmillan Distribution Ltd, Houndmills, Basingstoke, Hampshire RG21 6XS, England

---

Nineteenth-Century British Secularism

Science, Religion, and Literature

Michael Rectenwald
Scholar, Liberal Studies, New York University, USA

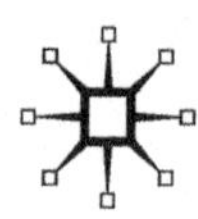

Softcover reprint of the hardcover 1st edition 2016 978-1-137-46388-3

First published 2016 by
PALGRAVE MACMILLAN

Palgrave Macmillan in the UK is an imprint of Macmillan Publishers Limited, registered in England, company number 785998, of Houndmills, Basingstoke, Hampshire RG21 6XS.

Palgrave Macmillan in the US is a division of Nature America, Inc., One New York Plaza, Suite 4500 New York, NY 10004-1562.

Palgrave Macmillan is the global academic imprint of the above companies and has companies and representatives throughout the world.

ISBN 978-1-349-69061-9
E-PDF ISBN: 978–1–137–46389–0
DOI: 10.1057/9781137463890

Library of Congress Cataloging-in-Publication Data is available from the Library of Congress

A catalog record for this book is available from the Library of Congress

A catalogue record for the book is available from the British Library

Typeset by MPS Limited, Chennai, India.

Contents

Acknowledgements vi

Introduction: Secularity or the Post-Secular Condition 1

1 Carlyle and Carlile: Late Romantic Skepticism and Early Radical Freethought 16

2 *Principles of Geology*: A Secular Fissure in Scientific Knowledge 45

3 Holyoake and Secularism: The Emergence of 'Positive' Freethought 71

4 Secularizing Science: Secularism and the Emergence of Scientific Naturalism 107

5 The Three Newmans: A Triumvirate of Secularity 135

6 George Eliot: The Secular Sublime, Post-Secularism, and 'Secularization' 168

Epilogue: Secularism as Modern Secularity 197

Notes 202

Index 248

Acknowledgements

This book owes a great deal to many colleagues, friends, and supporters who have helped guide its contents and encourage its making. For generous and timely readings of the proposal and/or parts of the manuscript, I am indebted to Patrick Corbeil, Anthony Galluzo, Nancy Henry, David Hoelscher, Stephanie McCarthy, Edward Radzivilovskiy, Dylan Rectenwald, Victoria Sams and Sarah Jo Skaggs. I am especially indebted to my long-time mentor, Jon Klancher. Jon's indispensable advice on the Introduction to this book represents what I take to be a major contribution.

I am grateful for the inspiration and support of the series editor, David Nash, and for the generous guidance and shepherding of this book through the publication process by Palgrave Macmillan editor Jenny McCall. I thank Jenny's editorial assistant, Jade Moulds, for keeping me on task and interceding on my behalf.

I owe a tremendous debt of gratitude to my long-time friend, confidante, and editorial and research assistant, Lori R. Price. Lori's tireless and ongoing readings of all parts of this manuscript in nearly every phase of its production have served not only to improve my writing but also to stimulate it. I cannot say enough about Lori's unflagging support, continual guidance, and steadfast belief that this book would come to fruition.

Finally, I want to thank my life-partner and the love of my life, Sarah Jo Skaggs. Thanks to Sarah, I have been endowed with the space, time, and deeply peaceful environs that I desperately needed to complete this work. Most importantly, her patience and love have been the mainstays of my life.

Parts of Chapters 3 and 4 of this book represent revisions of my essay, (2013) 'Secularism and the Cultures of Nineteenth-Century Scientific Naturalism', originally published in the *British Journal for the History of Science* 46.2, 231–54. I want to thank the editor and publisher for permission to reprint parts of this material.

I would like to thank the New York University Center for the Humanities for their support of this project.

Introduction: Secularity or the Post-Secular Condition

This book addresses the recent criticism and breakdown of the secularization thesis, a development that amounts to a crisis in the concept of secularism and in the long-held assumptions about an inevitable modernization from traditional, religious worlds to secular ones. Until the last decades of the twentieth century, secularization was generally regarded as a nearly indisputable fact of modern life and a staple of sociological thinking. A broadly held belief in secularization, what I call 'the standard secularization thesis', pointed to religion's continual and inevitable decline. In the conjectures of the earliest sociologists – including Auguste Comte, Karl Marx, and Max Weber – secularization was considered an inevitable result of modernization: urbanization, industrialization, the rise of science, individualization, and so forth. Secularization was understood as teleological and irreversible, ending in the ultimate extirpation of religion and 'the death of God'.[1] As an example of this article of faith, in 1968, the American sociologist Peter Berger was quoted in the *New York Times* as predicting that '[b]y the 21st century, religious believers are likely to be found only in small sects, huddled together to resist a worldwide secular culture'.[2]

Yet unanimity among scholars regarding the progress of secularization no longer subsists, to say the least. Erstwhile proponents of the secularization thesis, including Berger himself, have conceded the persistence and continued relevance of religion, which has proven to be much more durable than they had imagined.[3] The universality, timing, and mechanisms of the standard secularization thesis have come under the severe scrutiny of scholars from a number of fields, and some have even suggested that we abandon the notion altogether.[4] 'A triumphalist history of secularization', as Talal Asad poignantly dubs it, has yielded to heated debates over a number of models for how secularization occurs and what it might actually mean.[5] The contemporary moment has even taken on a new moniker: post-secular.[6] Indeed, while important thinkers have reasserted the secularization thesis, and others have attempted to retain it with significant revisions, there is little doubt that it has been significantly weakened.[7] Secularism and secularization, that is, are no longer regarded unquestionably as the vaunted pillars of modernity.[8]

Such challenges to secularism and the secularization thesis have not left historical work untouched, and this is especially the case in terms of nineteenth-century studies. Traditionally, the nineteenth century in Britain has represented a pride of place within the secularization narrative; 'the age of Darwin, the age of steam, the age of the first self-identified secularists' represented a watermark of secularization.[9] Romanticism was seen as a kind of aesthetic secularization or aestheticism as secularization, while a 'crisis of faith' narrative predominated in understandings of the Victorian period. The Romanticism as secularization paradigm pointed to the translation of traditional Christian religiosity into secular spirituality among Romantic-age writers and artists. The 'crisis of faith' narrative featured (mostly middle-class) Victorian intellectual heroes whose renunciations of religious creedal commitments signaled a progressive and teleological secularizing trend.[10]

Given the challenges to the standard secularization thesis, however, Romanticists have undertaken reassessments of the dominant motif.[11] And in Victorian studies, the new 'religious turn'[12] has even given rise to a countervailing narrative meant to replace the secularization thesis and the crisis of faith narrative, most emblematically dubbed the 'crisis of doubt' by Timothy Larsen.[13] Larsen's coinage is meant to suggest that the 'crisis of faith' in the period has been grossly overestimated, while doubt itself was in crisis, as many erstwhile Secularists doubted their doubt and reconverted to some form of Christianity. Along similar lines, Callum G. Brown refers to two paradigms that have been adduced for understanding the religiosity of nineteenth century Britain: the 'traditional, "pessimist" view of religion', under which religion declines invariably from the early nineteenth-century on; and 'a revisionist school of "optimist" scholarship ... which argued more directly that the theory of secularization was wrong in whole or in part because it failed to account for the observable success of religion in nineteenth-century British industrial society'.[14] According to Brown, both of these schools are mistaken – the prior because it posted secularization far too early, and the latter because it left the standard secularization thesis intact, while merely recalibrating it for religion's survival in the nineteenth century. Most problematically, '[s]cholars ... have been trying for years to qualify or disparage secularization theory on its own terms – using the same methods and the same conceptualisation of the issue'.[15] Meanwhile, the thesis itself, Brown argues, should be overthrown. According to Brown, secularization did not happen according to this model, but rather took place later, much more suddenly, and for different reasons than those given by the standard secularization thesis. Thus, Brown suggests that the secularization thesis has been a major impediment to understanding secularization.

While such antithetical paradigms describe what each holds to be the dominant trend in the period, both models miss a sense of just what secularization might mean and just how secularity might be characterized as such. Against both tendencies – and against Brown, who post-dates secularization after the mid-twentieth century, as well as against recent thinkers who claim that it never happened[16] – I answer the intriguing and important challenge effectively issued by David Nash by proposing a new paradigm that not only comprehends both secularization (or the crisis of faith) and the persistence of religiosity (or the crisis of doubt) but also that moves beyond the language of crisis altogether – or, as I see it, one that accounts for both while favoring neither, while also embracing a broad range of other options and predicaments.[17] Engaging critically with the notion of secularity put forth by Charles Taylor,[18] I heed David Nash's recent recommendation that historians of religion (and by implication, historians of secularism) 'look beyond the teleological straightjackets [of the secularization thesis] that previously restricted and encumbered them'.[19] *Nineteenth-Century British Secularism* offers a paradigm that obviates the adjudication between crisis of faith and crisis of doubt narratives (or between secularization and its lack). Instead, this volume figures both the crisis of faith and the crisis of doubt in terms of a new understanding of an emergent secularity, as emblematized in particular by mid-century Secularism proper. I will address the crisis in the secularization thesis by foregrounding a nineteenth-century development called 'Secularism' – the particular movement and creed founded by George Jacob Holyoake from 1851 to 1852 – in connection with several other secular interventions in the nineteenth century and as an instantiation of the rise of modern secularity. While Secularism proper has been treated by historians and other scholars – having been examined in terms of social history,[20] literary studies,[21] feminism,[22] and even the history of science[23] – it has yet to be situated in terms of so-called secular modernity, or especially in connection with the much-disputed processes of secularization. *Nineteenth-Century British*

Secularism rethinks and reevaluates the significance of Secularism, regarding it as a distinct historic moment of modernity and granting it centrality as both a herald and an exemplar for a new understanding of modern secularity, and as an inaugural event of the post-secular condition.

I have made mention of a number of distinct yet similar terms, so I shall briefly define them here, contrasting when necessary their meanings during the period with their employment today, while explaining how they are defined and mobilized in this book. This segue will also amplify my arguments regarding the secular, Secularism, secularization, post-secularism, and secularity.

Secular: In the nineteenth century, the word 'secular' referred, as it does today, to the non-religious. But it also signified the worldly aspects of 'this life'; that is, it gestured toward the concerns of existence on earth as opposed to eternity or another world, and to the activities for maintaining and living an earthly life as opposed to the aspirations of religious life or spiritual improvement. Thus in *The Missionary Magazine* for 17 March 1800, in a life of John Bunyan, the beginning of Bunyan's religious conversion is described as follows:

> Such an entire change took place in his sentiments, dispositions, and affections, and his mind was so deeply engaged in contemplating the great concerns of eternity, and the things pertaining to the kingdom of God, that he found it very difficult to employ his thoughts on any secular affairs.[24]

'Secular affairs' signified those duties or activities involving other than spiritual, otherworldly concerns of existence, specifically in this case those pertaining to earning a living, or 'keeping body and soul together'.

As the above passage suggests, the word 'secular' was originally contrasted not to religion, but to eternity. Derived from the Latin, *saeculum*, the secular is related to time, and the French word for century, *siècle*. The secular thus stood for occurrences in worldly time as opposed to otherworldly eternity, to temporal as opposed to spiritual existence.

From the late thirteenth century, the secular came to refer to members of the (Roman Catholic) clergy who lived outside of monastic seclusion and served the laity. The secular clergy were contrasted with the cloistered monks and were generally considered less religiously rigorous than the latter. This in fact was the first meaning of 'secular' in connection with religion, and although this sense can occasionally be found in use, the term had generally lost this signification by the middle of the nineteenth century. It is important to note that the 'secular' in this sense was a term within religious discourse. Then, the secular, which had once denoted a lesser state of religiosity within Christianity, later came to mean anything that was outside of religious observance or practice altogether. In the Secular movement beginning in the mid-nineteenth century, the secular represented that which pertained to 'this life' as opposed to another, and the means for the improvement of 'this life'. Such means were generally termed 'Science', deemed the sole 'Providence' of humanity.

In addition to the uses made of it in the period, I employ the term 'secular' as part of the secular-religious binary, a binary that is troubled in various contexts and to varying degrees, but where the secular generally indicates the non-religious. But the secular should not be understood as the mere absence of religion as such. One of the arguments of *Nineteenth-Century British Secularism* is that the secular, far from being merely a space devoid of religion, is never neutral or content-free; rather, the secular always contains substantive elements, including social, political, economic and other content and meaning.[25] Further, the content of the secular is always context-dependent, and the secular's emergence, rather than being an inevitable result of 'progress', the ineluctable march of history, or the outcome of a progressive, irreversible, teleological secularization itself, is always contingent and subject to local conditions. The secular arises in response to and as a vehicle for authority and contest within particular circumstances. Following Asad's assertion that the secular 'is neither singular in origin nor stable in its historical identity',[26] this book provides accounts of the nineteenth-century emergence of the 'secular' – in various public spaces, discourses, and practices, including science, religion, literature, and social and political movements. Additionally, in *Nineteenth-Century British Secularism*, the secular and the religious are regarded as mutually co-constitutive; they derive their substance and meaning only in distinction from one another.[27] And, as David Nash suggests, the secular and the religious are often found operating to similar effect, as the same narratives may be deployed by secular and religious culture.[28] Thus, the secular is not necessarily the negation of the religious and the secular and the religious are not necessarily antinomies.

Secularism: The word was coined by George Holyoake and first used by him in his periodical the *Reasoner* on 25 June 1851.[29] As distinct from its contemporary connotations, the neologism as first mobilized referred not to any general prevalence of the secular in the state or the public sphere, or to the absence of religion as such. Rather, 'Secularism' was invented *as a substitute for atheism* – to refer to 'the work we have always had in hand' in the freethought movement, which Holyoake and company were in the process of reconstituting and which reconstitution was to be marked by the new term. Secularism referred specifically to a developing 'positive' freethought movement and creed, and stood for new ecumenism embracing both secular and religious elements and participants. I treat Secularism proper throughout, but most directly in chapters 3, 4, 5 and the epilogue.[30]

Secularization: As I have suggested above, this is perhaps the most complicated and controversial of the terms. First referring to the transfer of church property to the state or private landholders during the Protestant Reformation, for example the expropriation and enclosure of monastic property under Henry VIII in England, later usage was extended to designate any transference of authority from religious persons or institutions to persons or institutions with non-religious functions. In contemporary parlance, secularization has often signified the (supposedly progressive and unidirectional) decline in importance and influence of religion in public life and private conviction, and has become nearly synonymous with the process of modernization itself. As I have suggested, the shape, extent, and even the very reality of secularization has been called into question over the past thirty-plus years, while many revisions of the secularization thesis have been proffered. My uses of the term secularization follow these contemporary understandings, but this book intervenes in the contemporary debates regarding secularization in ways that I have alluded to above, and discuss further below.

Post-secularism: An ambiguous and contested term, post-secularism may signify a skepticism or antagonism toward secularism in recognition of the persistence or 'resurgence' of religion. Connected with post-colonialism, post-secularism may regard secularism as a legacy of colonialist enterprises and a disguise for the domination of a particular (Christian) religious order. Regarded in connection with postmodernism, in which Jean-François Lyotard and others call into question the self-arrogating proclamations of a progressive and teleological modernity, post-secularism poses a challenge to secularization as a master narrative.[31] By post-secularism, I refer to 'an attempt to overcome the antinomy of secularism/religion',[32] such that both are granted recognition under a common umbrella. Post-secularism recognizes the persistence of religion and marks an acknowledgement of a religious and secular *pluralism*. Post-secularism accords to religion an enduring value – a place at the table in politics, a voice in the public sphere, and an abiding role in private life. It recognizes the ethical resources and community-building efficacy that religious bodies and their ministers can offer and countenances the function of religion in constructing and defending cultural identities. Further, by acknowledging and respecting the persistence of religion, post-secularism amounts to a refutation of the standard secularization thesis. According to post-secularism, the secularization thesis has been empirically disproven. Rather than a descriptive characterization of modernity, the secularization thesis, post-secularism suggests, is a normative imperative and a (failed) self-fulfilling prophecy of secular advocates. As Aleksandr Morozov puts it:

> 'Secularisation' as an all-embracing process no longer exists, but the reason it no longer exists is not because it has come to an end as a process with the onset of the postsecular age. The reason is rather that there never was such a process. There was only self-description on the part of the rationalising consciousness, which singled out this process as real and significant.[33]

Secularization, if it has indeed happened, has not followed the patterns set out by the standard secularization thesis but rather has resulted in something like the post-secular condition, or what I refer to throughout this book as modern secularity: the continued co-existence and mutual reproduction of the religious and the secular by its Other.

Secularity generally refers to the condition of being in a (more or less) secular society as such. However, as I mobilize it in this book, secularity borrows something from Charles Taylor's definition in *A Secular Age*. After noting the usual meanings of secularity as 1) the expulsion of religion from sphere after sphere of public life, and 2) the decline of religious belief and practice, Taylor defines 'secularity 3' as follows:

> Secularity in this sense is a matter of the whole context of understanding in which our moral, spiritual or religious experience and search takes place. By 'context of understanding' here, I mean both matters that will probably have been explicitly formulated by almost everyone, such as the plurality of options, and some which form the implicit, largely unfocussed background of this experience and search, its 'pre-ontology', to use a Heideggerian term.[34]

Leaving Taylor's controversial philosophical historiography aside, with this sense of secularity, as I understand it, Taylor seems to suggest a new understanding of what it means to live in a secular age, and a different understanding of the relationship between the secular and the religious in that age. Rather than positing the antinomy of the secular and religious, the term secularity is deployed to describe an abiding tensile condition comprising the coexistence of the religious and the secular within a common frame.

Secularity as it concerns belief amounts to what I call in this volume an overarching *optative condition* – which comprehends the various possibilities of belief and unbelief as well as the irresolution, tensions, and continuing challenges that they pose to one another. Under this notion of secularity, the persistence of religion is acknowledged, but, as Taylor notes, religion has become a choice among other options. But, it is also a condition under which the very structure of belief may have been changed. Under modern secularity, religiosity has been altered by the secular and relativized as one possibility among others, a relativism that profoundly impacts and disrupts it. Religious belief, where it survives, is inevitably contingent and unstable. Thus, this conception of secularity theoretically accounts for the fragility and vacillations of religious belief and unbelief, perhaps even making sense of the putative post-secular 'religious resurgence' observed by Peter Berger.[35]

So, why is it important to recuperate and feature the version of Secularism that George Holyoake founded in mid-nineteenth century Britain? First, because it arises from what might be thought of as an unexpected social provenance – not a world of elite intellectuals with their highbrow periodicals like the *Westminster Review*, but rather from the periodical and publishing houses of artisanal and working-class political activists, leaders, and journalists struggling for political representation, the rights of 'free' expression, and economic and political autonomy. Working- and artisanal-class freethinkers had promoted irreligious positions decades in advance of middle-class skeptics and unbelievers in nineteenth-century Britain. It is no surprise then that they arrived at the notion of Secularism before middle-class thinkers (although, as we shall see, not without the latter's help). Secondly, the movement shows how Secularism was a contingent, historically shaped mode of action that could have turned out otherwise. Its contingent character challenges any extant notion of secularism as a universal doctrine delivered wholesale by Enlightenment rationality on the doorstep of the nineteenth century. (This fact also enlarges our understanding about the contingent and plural character of *contemporary*, context-dependent and local secularisms; they are not anomalies but rather have a precedent in western historical Secularism itself; there never was a (logically necessary) secularism; there were always only possible secularisms.) Third, Secularism as founded by Holyoake illustrates the way Secularism as a real-world movement already responded to the failures of Enlightenment rationality to replace religion by admitting to the abiding presence of religion and welcoming the religious believer to its fold. Significantly, Secularism as first developed was never strictly an atheism or antitheism. To the contrary, developed explicitly as an alternative to atheistic freethought, Holyoake's Secularism anticipated Thomas Huxley's agnosticism by nearly two decades. Thus Norman Vance is mistaken in conflating Holyoake and Charles Bradlaugh as two militantly 'anti-religious Victorian freethinkers'.[36] Finally, given its inclusion of religious discourse and practice, Secularism anticipated the post-secular moment announced in the early twenty-first century and debated amongst scholars of secularity today. The development of mid-nineteenth-century Secularism proper demonstrates that, as Rajeev Bhargava puts it in a related context, 'we have always been post-secular'.[37]

In addition to Secularism proper, the book treats several important secular interventions in the nineteenth century, including Thomas Carlyle's 'natural supernaturalism', Richard Carlile's anti-theist science advocacy, Charles Lyell's uniformity principle in geology, the mid-century emergence of scientific naturalism, Francis Newman's naturalized religion or 'primitive Christianity', and George Eliot's secularism and post-secularism. Some of these figures, such as Newman, Holyoake, and even Eliot, were more or less directly involved in the development of Secularism proper. Others, such as Carlile, Carlyle, and Lyell, contributed to the underlying episteme from which Secularism proper evolved. Taken together, they contribute to an important cultural, philosophical, political, religious and scientific current whose repercussions would be felt throughout the nineteenth century and beyond.

These illustrations of secularity by no means constitute a comprehensive account of what has been called the 'secularization' of British society – mostly because this is not the picture I am drawing or the model under which I am operating. Following the commencement of the 'reshaping of religious history' announced by David Nash, this history of secularism/Secularism does not begin with the assumption of a secular teleology.[38] And, while more than mere tokens of the secular, these instances of secular emergence are meant as epitomes rather than the pieces of a complete puzzle. Thus, for example, while I touch on Darwinism throughout, other than treating it in terms of its connection to the emergence of scientific naturalism (Chapter 4), and the Anglo-Jewish response (Chapter 6), the Darwinian revolution is largely left unexplored in these pages. The reasons for this apparently glaring omission are several. First, the secular had already emerged decades ahead of the publication of the *Origin of Species*. Even considering only the so-called middle-class Victorian 'crisis of faith' phenomenon, we note that evidence for such a crisis exists as early as 1840, if we take Charles Hennell's *An Inquiry Concerning the Origin of Christianity* (1838) as a somewhat arbitrary marker. But by the 1840s, the effects of the biblical Higher Criticism were already being felt by those who, like Mary Ann Evans (George Eliot), were exposed to it (and, in her case, exposing others to it).

Further, as I show in Chapter 2, even within the milieu of gentlemanly geology, the secular made inroads in science by 1830 with the publication of Charles Lyell's *Principles of Geology* (1830–1833). And in other milieus, plebeian propagandists for a materialist science not only produced and disseminated evolutionary ideas well before 1859 but also they fleshed out the implications of such theories in terms of the secular well in advance of the watershed publication event of 1859. Rather than fearing a loss of religious faith or experiencing it as catastrophic, these artisan radicals gladly embraced materialist cosmologies and advocated doctrines that supported their anti-clerical, republican, and radically egalitarian worldviews. Therefore, while it may be true, as Robert M. Young suggested decades ago, that Darwinism did not precipitate a major gestalt shift in the period,[39] it is also the case that materialist cosmologies, historicist biblical criticism, and geological science had already begun to irrupt decades before the *Origin*, whether or not these intellectual episodes registered any significant sociological effect. In any case, a presupposition of the Darwinian 'origin story' of secularization is that science is necessarily a secular and secularizing force, and thus with the publication of such texts as the *Origin*, a secularization process is inevitably put into play. This study interrupts this assumption by showing that science is far from necessarily secular or secularizing and that rather than necessarily precipitating the secular, science itself must be *made* secular before it is to have any such secularizing effect. I treat the emergence of the secular in science in chapters 2 and 4, showing its contingent and context-dependent character as opposed to its supposed 'natural' secularism as such. Further, to presume that such revolutionary science naturally unsettles religious belief is to accord it an efficacy that it does not necessarily have, especially when considering that earlier scientific revolutions in conflict with Biblical narratives did not precipitate faith-shattering consequences but rather were accommodated rather well by traditional Christianity. The Copernican revolution is a striking case in point. Milton's *Paradise Lost* (1667), for example, easily accommodated the Copernican cosmology without evincing a loss of Christian belief. As Charles LaPorte puts it, 'to take modern science as categorically inimical to religious belief is to misread most of modern history'.[40] This is not to say that the Darwinian revolution or Lyell's geological science did not result in repercussions, but it is to acknowledge that the impact of paradigm-shifting science on belief is contingent upon social contexts, contexts that are explored in chapters 2 and 4.

Another apparent omission is the Bradlaugh branch of Secularism centered on the National Secular Society (NSS) and the *National Reformer*, the periodical founded and co-edited by Charles Bradlaugh to advance a secular agenda. Although I treat Bradlaughian Secularism in chapters 3 and 4, Bradlaugh and company are dealt with specifically as they interact with and differ from the Holyoake branch of Secularism. The reasons for this emphasis will be made clear, but I will note here that mid-century Secularism as founded and developed by Holyoake is the central object of interest in these pages. This interest has to do with my argument that Holyoake's brand of Secularism represents an inaugural event in modern secularity and an anticipation of the post-secular.

Doubtless other important phenomena would appear to be necessary in order to register a complete map of the emergence of the secular in the period. Robert Owen and Owenism are not directly treated, although I pick up the legacy of Owenism with George Holyoake's Secularism, which is generally understood to be the successor to Owenism. The British biblical criticism, in particular the publication of *Essays and Reviews* in 1860, is certainly another. While I do treat the effects on British thought of German Higher Criticism in Chapter 5, and also the Anglo-Jewish response to the Higher Criticism in Chapter 6, my method is not one of 'coverage' so much as illustration of the notion of secularity being proffered, which the following chapters describe.

Chapter 1 deals with two antithetical figures – Thomas Carlyle and Richard Carlile – whose greatest similarity may be their homonymic surnames. This chapter shows how Carlyle and Carlile represent and propose differing versions of secularization, thus exemplifying the notion of secularity that I am employing throughout. At first blush, these two figures could not be any further apart philosophically, and yet they are bookends of the secular as it emerges in the period. Richard Carlile's freethinking career uncannily epitomizes the rationalism and utilitarianism that Thomas Carlyle lambasted repeatedly – particularly in 'Signs of the Times' (1829), 'Characteristics' (1831), and *Sartor Resartus* (1831). Whereas Carlyle's *Sartor Resartus* represents a Romantic re-enchantment of the secular and 'immanentization' of the divine, Carlile's *Address to*

Men of Science (1821) embodies the mobilization of a hard secularism in an attempt to eradicate all semblances of religiosity. Ironically, Carlile would express this anti-religious, anti-theist desire in millennial, evangelical tones. Together, Carlyle and Carlile stand for two tendencies of the secular in the period. They adumbrate the coming of Secularism as it would emerge by mid-century but also they are figures fully immersed in a new secularity: a condition embracing belief, unbelief, and a suspension between the two.

In Chapter 2, I treat the field of gentlemanly geology during a period of a great explosion in knowledge production in order to show the contingent and context-dependent character of the emergence of the secular in science. Charles Lyell – in connection with the Murray publishing house and the Tory *Quarterly Review*, a bastion of political and religious conservatism – called for a reform of science and educational institutions based on the dramatic upsurge in scientific activity underway from the early nineteenth century. Lyell's scientific knowledge project can be seen largely as a response to such plebeian educational plans and projects as promoted by Carlile in his *Address to Men of Science* and the Zetetic societies modeled after it, the numerous Mechanic's Institutes founded thereafter, and the projects inaugurated by the Society for the Diffusion of Useful Knowledge (founded in 1826). Furthermore, like Carlile's proposals and projects, Lyell's knowledge project, which included his *Principles of Geology* (1830–1833), may be understood as secular. The project was aimed at the supersession or circumventing of theological and other cultural strictures within the domains of knowledge production and dissemination. It depended upon the *differentiation* of spheres – scientific, educational, and to some extent the broader public sphere – and the clearing of spaces within them to make room for new configurations and understandings of science and education. The chapter shows how a conservative publisher and a progressive author worked together to advance a secular, reformist agenda in a gentlemanly milieu of scientific knowledge production.

Chapter 3 turns to the movement of Secularism founded by Holyoake from 1851–1852, tracing the shift in freethought from the negation of theism to a 'positive' new movement and creed independent of, but not necessarily opposed to, religious belief. The chapter develops the history of Holyoake's Secularism in connection with several trends in the period; first, the break-up of the older infidelity represented by Richard Carlile in the 1820s and continued through the 1840s; second, as a development and differentiation from Robert Owen's social environmentalism and cooperation movement; third, a movement toward a broadened inter-class, cross-belief affiliation, represented particularly in the association (a 'Confidential Combination') of Holyoake and company with such figures as Thornton Hunt, George Henry Lewes, George Eliot, Herbert Spencer, Francis Newman, and Thomas Huxley; and finally, the eventual divergence from Holyoake's brand by the later Secularist strain headed by Charles Bradlaugh, especially in terms of the issues of atheism, sexuality, and birth control. With Secularism, Holyoake developed a big tent movement under which theists, unbelievers, and skeptics could combine for the material improvement of humanity, especially the working classes, using 'science', broadly conceived, as their method. Mid-century Secularism, I argue, should be understood as a salient moment of modernity, marking as it does an inaugural expression of modern secularity understood as defined above. That is, with Secularism, Holyoake was already engaging in a post-Enlightenment notion of secularity as a pluralistic, inclusive, and contingently constructed combination of believers and unbelievers. Within a state that had only recently criminally persecuted blasphemy, with himself as the state's most recent victim, Holyoake nevertheless already grasped a sense of secularity as characterized by the recognition and cooperation between religion and its others, a vision of the public and political spheres not unlike that which Habermas describes as 'post-secular'.

Chapter 4 examines the importance of Holyoake's brand of Secularism to the creed of scientific naturalism – the epistemological creed that supported and promoted Darwinism, as developed and promoted by Thomas H. Huxley, John Tyndall, Herbert Spencer, and others. Drawing on a philosophical family resemblance and evidence of extensive social contact, I argue that Secularism was a significant source for the emerging new creed of scientific naturalism in the mid-nineteenth century. Not only did Holyoake's Secularism help clear the way for scientific naturalism by fighting battles with the state and religious interlocutors but also it served as a source for what Huxley, almost twenty years later, termed 'agnosticism'. As I show in Chapter 3, Holyoake modified freethought in the early 1850s, as he forged connections with middle-class literary radicals and budding scientific naturalists, some of whom met in a 'Confidential Combination' of freethinkers. Secularism became the new creed for this coterie. As I show in this chapter, Secularism promoted and received reciprocal support from the most prominent group of scientific naturalists, as Holyoake used Bradlaugh's atheism and neo-Malthusianism as a foil, forging and maintaining friendly relations with Huxley, Spencer, and Tyndall through to the end of the century.

In Holyoake's Secularism, I argue, we find the beginnings of the mutation of radical infidelity into the respectability necessary for the acceptance of scientific naturalism, and also the distancing of later forms of infidelity incompatible with it. Holyoake's Secularism represents an important early stage of scientific naturalism, and scientific naturalism marks an important moment in modern secularity. But perhaps more importantly, as I have suggested above, Secularism's role in the emergence of scientific naturalism underscores the contingent relationship between science and the secular. Science is not necessarily secular as such; as this chapter shows, it has to be made secular.

Chapter 5 examines the impact of secularism/Secularism on religious discourse, and vice versa. It registers a watermark in modern secularity – showing that the secular is not merely a space separate and distinct from religion, but rather that it infiltrates and conditions religion itself. The chapter treats the three Newman brothers – (Cardinal) John Henry, Charles Robert, and Francis William. Beginning from the same evangelical and familial base, these three Newmans diverged toward three different belief destinations: Catholicism, atheism, and theism. They thus illustrate secularity beautifully. I pay particular attention to Francis Newman, the liberal theologian and advocate of secular improvement. Francis Newman is a pivotal figure for Secularism/secularism in the period, especially given his impact on Holyoake and Darwin. I examine Newman's religious works, especially *The Soul: its Sorrows and Aspirations: The Natural History of the Soul, as the True Basis of Theology* (1849) and *Phases of Faith: or, Passages from the History of My Creed* (1850). These treatises stand as milestones for the secularist impulse in mid-century religious discourse – and as widely divergent from the Catholic revival undertaken by his brother in the Tractarian (or Oxford) Movement. I argue, however, that both moves – Francis Newman's naturalization of religion and 'immanentization' of God, and John Henry Newman's Catholic revival – are driven by the same condition of secularity, in particular the challenges posed by rationalism for evangelical Christianity from the end of the first quarter of the century. Of the three brothers, Francis Newman best represents the condition of secularity, taking as he did a middle course between the orthodox Christianity of John Henry and the unbelief of Charles Robert. This chapter shows how religious discourse was impacted by the secular but also how Secularism proper was constructed in conversation with this new religiosity as represented by Francis Newman.

Chapter 6 examines the literary representations of religion and secularism in the fiction of George Eliot, paying particular attention to her final novel, *Daniel Deronda* (1876). The mid-century 'crisis of faith', secularism, and 'secularization' have generally been treated almost exclusively in connection with Christianity; to redress this remission, this chapter turns to examine 'secularization' in the context of Judaism, first as represented in Eliot's fiction, and then briefly in Great Britain in the last quarter of the century. Eliot was a committed secularist. However, I argue that given the recognition that she lent religion and the importance that she placed upon it within a secular framework, she is best regarded as a 'post-secularist'. Although her earlier fiction generally repurposed religiosity for secular ends, in *Daniel Deronda,* Eliot takes a surprising and ambiguous 'religious turn'. *Daniel Deronda* represents a secular-religious novel that accords greater importance and centrality to religion, in particular to Judaism. I consider Judaism in connection with the Eliot's use of the trope 'blood', to examine whether this figure stands for 'racial' determinism or cultural inheritance, which bears significance in terms of Judaism's apparent exceptionality. After a discussion of difference and transcendence in *Daniel Deronda,* I consider the question of 'secularization' in connection with nineteenth-century Anglo-Judaism, a line of inquiry that has been largely neglected, and one that I aim to inaugurate with this chapter.

Finally, in the epilogue, I explore a central tension within Secularism, a tension which continues to play out to this day, and which can be seen even in contemporary frameworks like post-secularism. But again, I suggest that this historical and contemporary tension may be explained in terms of the notion of secularity that I investigate throughout. Modern secularity or the post-secular condition simply mirrors the same tension that Holyoake's Secularism embodied over one hundred and sixty years ago.

1
Carlyle and Carlile: Late Romantic Skepticism and Early Radical Freethought

As I have suggested, mid-century Secularism as founded by George Jacob Holyoake in 1851–1852 was a post-Enlightenment development, both an extension of Enlightenment rationality, and a response to its failed promises for extending reason across the public and private spheres to the exclusion of religious belief and practice. In order to comprehend this development, I begin by examining some salient post-Enlightenment discourse and activity in early nineteenth-century Britain. This chapter counter-poses two exemplary, late Romantic-age and seemingly antithetical successors to the Enlightenment legacy. One epitomizes the late Romantic response to what Romantics deemed an overweening faith in Enlightenment rationality, as expressed in terms of scientific materialism, Political Economy, and a Utilitarian ethical 'calculus'. The other represents the extension of Enlightenment promises to the 'popular Enlightenment' and the expression given it in the artisanal freethought movement, a movement that would eventually lead, circuitously, to Secularism proper. Respectively, the two figures – the 'Victorian sage' and cultural critic Thomas Carlyle and the Romantic-age, plebeian, Paineite radical, Richard Carlile – will serve to represent these currents. While apparently diametrically opposed, the standpoints of Carlyle and Carlile demonstrate a range of secular possibilities in the period.

The choice of Carlyle and Carlile may seem to be based arbitrarily on their homonymic surnames, but together, these two contemporaneous figures work well to frame the outer edges of the secular as I define it. I regard the secular not as the outcome of progressive religious decline – per the standard secularization thesis – although this sense of the secular is discussed throughout this book. Rather, I understand the secular as an element within secularity, an overarching or background *condition,* a new 'naïve framework' of modernity that embraces belief and unbelief, the secular and the religious, as well as the irresolution and challenges posed by the conjunction of these elements.[1] According to Charles Taylor, secularity is a 'modern imaginary' that, by the nineteenth century, involved all subjects in a new set of dilemmas and choices, which constitute what I am calling an *optative condition*. The development of secularity precedes the period under consideration, but by the nineteenth century, secularity develops into what Taylor refers to as a 'nova', as its contours become spectacular by virtue of the diversity that it permits. As Taylor notes:

> the salient feature of the modern cosmic imaginary is not that it fostered materialism, or enabled people to return as it were to religion, though it has done both these things. But the most important fact about it which is relevant to our enquiry here is that it has opened a space in which people can wander between and around all these options without having to land clearly and definitively in any one.[2]

Between them, Carlile and Carlyle represent a range of this *wandering* in the early nineteenth century – from religious faith, to skepticism, to materialism, to 'natural supernaturalism', to 'rational Christianity'. The metaphysical belief commitments that they present are also connected to 'worldly' convictions. Furthermore, both of these figures construe their choices as conditioned and constrained by the contexts that make them possible. Despite or perhaps because of their significant differences, Carlyle and Carlile illustrate the outlines of secularity that I am engaging here and throughout. Their views also illustrate theories of secularization itself – both the standard secularization thesis, as well as revised versions of secularization.

Thomas Carlyle's 'natural supernaturalism', from *Sartor Resartus* (1833–1834), has been taken by critics to represent a characteristic expression of Romantic secularization, placing 'belief' on a new naturalistic basis (albeit at the same time spiritualizing belief). On the other hand, Richard Carlile's early freethinking career may be seen as uncannily epitomizing the rationalism and Utilitarianism that Thomas Carlyle lambasted repeatedly, especially in 'Signs of the Times' (1829) and 'Characteristics' (1831). In his radical periodical and pamphleteering career, Carlile advocated the immediate secularization of the social order in its various domains. With a faith in science as an unmediated means of access to the phenomenal world available for social and political change, Carlile's scientism was a proto-positivism, embodying a progressive and teleological model of a declining religiosity. Whereas Carlyle represented the expression of the secular in religious terms (or vice versa), in his efforts to extirpate belief, Carlile advanced an emergent 'hard naturalism'.[3] On the other hand, Carlyle attempted to retain the higher purpose and meaning making potentiated by Christianity, while eliminating its doctrinal and miraculous basis (what he called its 'Mythus'). Carlyle and Carlile thus adumbrate Secularism proper, as it would emerge by mid-century. They represent antipodal figures, who are nevertheless immersed in a new common secularity.

Generally, in the nineteenth century in Britain, the religious and secular choices and dilemmas availed have been thought to include, broadly considered, established and dissenting Christianity; an evangelicalism that spanned the two; Unitarianism and other forms of theism and deism; Romantic reconfigurations of Christianity; pantheism; atheism; and later, secularism, agnosticism, rationalism, spiritualism, theosophy, and others. However, until relatively recently, the historiography of the period has been dominated by the familiar 'crisis of faith' narrative, a narrative that runs parallel to and reinforces the standard secularization thesis. Emboldened by challenges to the standard secularization thesis in broader histories and sociological studies, historians studying the nineteenth century have begun to challenge this dominant motif. One salient work, Timothy Larsen's *Crisis of Doubt* (2006), is especially relevant to this discussion.[4] In a critical intervention into the histories of freethought, secularism, and religion, Larsen coins the phrase 'crisis of doubt' to cleverly destabilize this dominant narrative. Larsen argues that contrary to the assumption of religious decline that has been vastly overplayed in historiography of the nineteenth-century, thriving religious belief was actually the rule, not the exception. To counter a long-standing preponderance of 'crisis of faith' historicism, Larsen conveys a series of reconversion, 'crisis of doubt' case studies, suppressed or lesser-known accounts of erstwhile Secularists, who later reconverted to some form of Christianity. Based on an opening critique of a broad body of scholarship, in conjunction with his collection of short religious re-conversion biographies, Larsen aims to overthrow the dominant versions of faith, doubt, and secularization that he sees as having distorted our perspective.

Like other relatively recent studies, such as Alister E. McGrath's *Twilight of Atheism* (2004), which disrupt the supposed inevitability of secular modernity, Larsen does well to point to the persistence and viability of religion in the period. He is also careful to acknowledge that the 'crisis of faith' really did happen for a number of subjects. However, in place of one stale and reductionist model, Larsen posits a competing hegemony, which leaves too little room for doubt, and makes faith rather too secure. Such a dichotomization, as either of these dueling and rather static, near all-or-nothing, faith or doubt paradigms suggest, belies the actual religious and secular diversity evident in the period. Likewise, rather than having to declare faith or doubt the ultimate victor, we might instead pay attention to the wide range of belief and unbelief commitments availed by nineteenth century circumstances.[5] We should understand secularity not only as embracing the 'crisis of faith' and the 'crisis of doubt' paradigms, but also as accounting for an increasing plurality of belief modalities available along a spectrum between the antipodes of faith and doubt, which were rarely static or fixed positions in any case. Further, such metaphysical commitments necessarily intersected with other convictions, including economic, moral, political, scientific, social and spiritual positions. This chapter begins an exploration of the kinds of belief commitments that modern secularity availed.

Natural supernaturalism: the 'desecularization' of the secular

A liminal text residing on the border between Romantic and Victorian literature and sensibility, *Sartor Resartus* has been treated as an instance of Romantic secularization as well as a prototype of the Victorian 'crisis of faith' narrative. In his *Natural Supernaturalism* (1971), M. H. Abrams considered the peculiar literary production in terms of the former, arguing that Romanticism itself was 'the secularization of inherited theological ideas and ways of thinking', and that the natural supernaturalism of *Sartor Resartus*, from which Abrams derived his title, represented the general tendency in the period 'to naturalize the supernatural and to humanize the divine'.[6] That is, for Abrams, the natural supernaturalism of *Sartor Resartus* was precisely the secularization of belief, the transformation of religious sentiment into a secular mode, a transformation triggered by the incursion of Enlightenment rationality, notably in the form of Utilitarianism and Political Economy.

Within the past two or three decades, as the standard secularization thesis has been challenged, studies in Romanticism have also undertaken a decoupling of Romanticism and secularization. As Colin Jager has noted, the Romanticism as secularization thesis has been challenged by studies that show religious belief to have been more important for canonical writers than suggested by critics such as Abrams.[7] This is clear in the cases of Wordsworth and Coleridge, as in Blake. The secularization thesis of Romanticism has also been contested by studies that point to a range of expression having little or no relation to secularization or religion. 'As a result, one might look to non- or extra-canonical writers and materials, and thereby contest secularization by, as it were, changing the subject'. Or, one might examine secularism in terms of its 'institutional dimensions', the conditions that make secularism possible or necessary.[8]

Along similar lines, Frank M. Turner has suggested that the 'crisis of faith' narrative – largely based on intellectual encounters, while prominent in the Victorian period and certainly applicable to the lives and works of several literary and philosophical figures – is otherwise an inadequate explanation for the emergence of the secular in the nineteenth century. Turner argues that religious discourse and particularly a new religious *pluralism* was equally or perhaps more important than secular literature. With the diversification of belief in the early nineteenth century, more opportunities for falling out with one's beliefs became possible, Turner suggests.[9] This position corresponds with sociologist Peter Berger's earlier claim that religious pluralism '*ipso facto* plunges religion into a crisis of credibility'.[10] This claim seems to be borne out by the number of defections from evangelicalism, for example. Historians and literary scholars have generally ignored this role for religion. Further, Turner argues that 'the widespread and widely accepted image of an existing religious faith ... that falls victim to emerging new intellectual forces' was born in the early nineteenth century and was largely owing to *Sartor*'s impact on subsequent writers and intellectuals.[11] *Sartor* forecasted a 'crisis of faith' made legendary by several prominent Victorian intellectuals. Indeed, famous Victorian 'crisis of faith' encounters – such as those of Alfred, Lord Tennyson, George Eliot, Leslie Stephen, Matthew Arnold, Francis W. Newman and others – may be read as variations on the *Sartor* theme, which itself mirrors an evangelical conversion.[12] While Turner may be correct in pointing to increasing religious diversity as a stimulus for secular conversions, his reading of *Sartor* is susceptible to the same tendency for which he criticizes historians. That is, much like Abrams, he reads *Sartor* as a straightforward secularization narrative wherein the secular merely displaces the religious. The religious has no real place in Turner's reading of *Sartor*; it is merely overthrown.

Certainly *Sartor* is a secularization allegory of sorts. As Barry V. Qualls has shown, the allegory reflects Carlyle's reworking of both the tradition of Christian pilgrimage, as popularized in Bunyan, and the Romantic secular rearticulations and re-locations of this tradition.[13] Within this allegory, moreover, the emergence of the secular does represent a crisis. The secular, as if an invading force, takes up residency within an exclusively religious sphere against a newly outmoded religious belief. The secular also involves disenchantment. The 'crisis of faith' is conditioned on freedom, but a freedom constrained in the context of an incontestable rationalism that has disenchanted the world – or, as Taylor maintains, has followed from the world's disenchantment.[14] Natural supernaturalism finally represents the possibility of faith in the providence of a benign and ultimately divinized nature. Yet the declaration of this faith is made necessary by secularity itself. In *Sartor*, belief and enchantment are not merely displaced by the secular, but are rather *reproduced* by it. Similarly, Joshua Landy and Michael Saler argue that secularization has always been accompanied not only by disenchantment, but also re-enchantment:

> Weber's account [of secularization] was, however, incomplete. What he neglected to mention is that each time religion reluctantly withdrew from a particular area of experience, a new, thoroughly secular strategy for re-enchantment cheerfully emerged to fill the void. The astonishing profusion and variety of such strategies is itself enchanting. Between them, philosophers, artists, architects, poets, stage magicians, and ordinary citizens made it possible to enjoy many of the benefits previously offered by faith, without having to subscribe to a creed; the progressive disenchantment of the world was thus accompanied, from the start and continually, by its progressive re-enchantment.[15]

Generally, the necessity to declare belief (or unbelief) is conditioned by secularity. The secular and the religious are mutually constitutive and substantive categories, inextricably wed. The faith that *is* declared in *Sartor*, while a post-confessional commitment, represents a re-enchantment of the secular, or a '*desecularization* of secularism'.[16] Natural supernaturalism follows from the dissolution of Christian faith and the necessity to live in a godless, material universe. Yet upon the collapse of an immersive Christian cosmology, the secular reproduces belief and re-enchantment. As a desecularization of the secular, this belief, natural supernaturalism, may be seen as secular, religious, or even as both. This is of less significance than the necessity to express belief, to re-enchant, at all. The 'Everlasting Yea' in *Sartor* is notable not merely as an affirmation of the supernaturalism of the natural, but critically, as a choice made possible or even necessary under the condition of secularity.

In *Sartor*, the secular is recharged, occupied by a purposive spiritual fullness of a kind previously experienced within Christian belief. The supernatural is immanent within the natural. But this is not the natural theology of Paley; nature is not to be read as a mechanical, semiotic system concocted by God for the decoding of His existence and goodness. With all its emphasis on metaphor and symbolism, natural supernaturalism does not finally rest on a reading of the 'Book of Nature' as text but rather represents the 'Word become Flesh' – in the cognitive and spiritual apprehension and recreation of nature by the seer – the apotheosis, not of nature *per se*, but of human-nature co-creation. In *Sartor*, this creative, co-productive apotheosis is the Romantic religion of art combined with what would become the Victorian ethic of duty.

In *Sartor Resartus* we are introduced to a (fictional) German philosopher "Diogenes Teufelsdröckh" by an unnamed English 'Editor', whose difficult mission it is to present the life and work of this extraordinarily strange and enigmatic German philosopher to an English-speaking readership.[17] Under this narrative conceit, the difficulties of the Editor's task include first and foremost Teufelsdröckh's German provenance, for Germany is 'one country where abstract Thought can still take shelter' amid 'the din and frenzy of Catholic Emancipations, and Rotten Boroughs, and Revolts of Paris, [which] deafen every French and every English ear'.[18] German philosophical idealism and Romantic mysticism, the Editor forewarns, may well challenge his own intellectual capabilities and knowledge background, while trying the patience, comprehension, and sympathies of his utilitarian-minded English readers.

The entirety of the Clothes Philosophy, the editor-narrator initially tells us, will consist of two parts, the 'Historical-Descriptive' and 'Philosophical-Speculative' (Books I and III of *Sartor* respectively).[19] To this is added, with the delivery of source material by Teufelsdröckh's apparent guardian and friend, Hofrath Heuschrecke, the biography of Teufelsdröckh (Book II of *Sartor*). The sources for Teufelsdröckh's biography are delivered in a set of six paper bags, each labeled with a zodiacal name and enclosing a welter of scrap papers, receipts, street advertisements, and other seemingly random parcels.[20] From these the Editor must construe the all-important biography – all-important because the philosophy of clothes is derived of experience rather than 'Logic'.[21] We need to know who the author really is in order to understand his new philosophy.[22] Thus, from all of the documents provided by Diogenes Teufelsdröckh, whose name translates as 'God-born Devil's-dung', the Editor composes *Sartor Resartus*, and introduces readers to his new Clothes Philosophy, a philosophy that is more astounding for having never been discovered before.

From the 'Preliminaries', we are reminded of the intellectual and social environment where the novel will intervene. In particular, the Editor remarks on the age's obsession with 'mechanism'. We encounter the names of Bichat, Lagrange, Laplace, Lawrence, Stewart, and others, who recall mechanical and materialist approaches to the universe so derided in Carlyle's earlier writing.[23] Teufelsdröckh's objections to such scientific approaches are elaborated later, especially in the 'Natural Supernaturalism' chapter of Book III, where he does not refute the sciences on their own terms, but nevertheless ruthlessly criticizes them for circumscribing knowledge within materialist limits. Here, the Editor continues, despite the concern of science with most everything, the greatest object, 'the grand Tissue of all Tissues, the only real Tissue', clothes, has not its devoted science. Thus the Editor figures the Clothes Philosophy as Teufelsdröckh's rejoinder to scientific materialism. Near the end of Book I, the Editor quotes Teufelsdröckh's summation of the historical-descriptive aspect of the philosophy. Clothing, or the outer appearance of all things, is the material world that enshrines the spiritual:

> It is written, the Heavens and the Earth shall fade away like a Vesture; which indeed they are: the Time-Vesture of the Eternal. Whatsoever sensibly exists, Whatsoever represents Spirit to Spirit, is properly a Clothing, a suit of Raiment, put on for a season, and to be laid off. Thus in this one pregnant subject of CLOTHES, rightly understood, is included all that men have thought, dreamed, done, and been: the whole External Universe and what it holds is but Clothing; and the essence of all Science lies in the PHILOSOPHY OF CLOTHES.[24]

The material outer garments are all-important as the only means of representation of spirit-to-spirit in time. And time is the only experience of the eternal, where everything nevertheless takes place:

> For man lives in Time, has his whole earthly being, endeavour, and destiny shaped for him by Time: only in the transitory Time-Symbol is the ever-motionless Eternity we stand on made manifest.[25]

Thus eternity is experienced only in time and not in a promised afterlife. Spirituality is a condition of earthly existence and is its own reward.

Yet the above passage also suggests a binary division of time and eternity. Indeed, binaries – such as those of time and eternity, matter and spirit, logic and experience, 'mensuration'/'numeration'[26] and wonder/imagination, the active and the passive, the mundane and the ethereal, the particular and the universal, the visible and the invisible, clothes and the body, the phantasmagorical and the real, phenomena and *noumena*,[27] and of course Devil's-dung and God-born – set up the distinction between the secular and the religious, the latter of which is not mere ritual or observance, but rather the *clothing* that virtually enshrines the spirit. Religion is the means by which is:

> invested the Divine Idea of the World with a sensible and practically active Body, so that it might dwell among them as a living and life-giving WORD ... These are unspeakably the most important of all the vestures and garnitures of Human Existence.[28]

That is, religion is the shroud necessary to make visible the invisible – to make the one 'Divine Idea' of the world at all perceptible. Thus, it is the secular that produces the visibility of the religious as such. One of the major arguments of *Sartor* is that religious vestures are time-contingent, not eternal. That is, religious vestures are also necessarily 'secular' (temporary, outer, changing) representations.

On the one hand, the secular is reinvested with, among other meanings, its oldest connotation – as concerned with long periods of *Time*: generations, centuries, and ages.[29] At its outer edge, Time bleeds into the religious in the '[t]he confluence of Time with Eternity'.[30] On the other hand, the secular splits in two; time also appears as the mundane, as the banal quotidian, particularly in the details of worldly life for which Teufelsdröckh is so ill suited. Teufelsdröckh's social alienation, and his cynical estrangement from the legal profession in the 'Getting Under Way' chapter of Book II (which ironically finds him doing nothing of the like, unless the rejection of his destined profession can be considered getting underway) signals the banality of lower-case secular time. In his search for a suitable course of action, a contrast is established between Teufelsdröckh's higher calling of contemplation, 'his Passive endowment',[31] and the means for warding off hunger, the active. The religious also splits in two; the internal or 'inward', which generally represents the religious,[32] is divided into the active and passive, the secular and religious. And 'Capability', or the great problem of matching *active* internal capabilities with *external* circumstances,[33] is a doubly secular matter, which nevertheless takes on religious significance; as we find in 'Centre of Indifference', the life of wonder is eventually constructed out of the 'Actual'.[34] Secular and religious elements are not only co-constitutive but also nearly indistinguishable and can even double as their Other.

In *Sartor*, we are confronted with an endless deferral and slippage of representation, such that the object of representation is seemingly never made manifest. Leaving aside the issue of the Editor's translation and narration, which adds a layer of mediation, even in the Editor's quotations of Teufelsdröckh, clothes represent underlying bodies, but bodies also are representations of something else; they are clothing. And while clothes serve as metaphorical representations of spirit, they also are a metaphor for this very representativeness. That is, in addition to being representations in their own right, clothes also represent the metaphorical, representational character of the text itself. In addition to the metaphorical, the text is also riddled with the symbolic; at least it claims to be: 'It is in and through Symbols that man, consciously or unconsciously, lives, works, and has his being', we read in the 'Symbols' chapter.[35] Ultimately, the referents for this plethora of metaphors and symbols appear to be unavailable. The text seems to suggest no means by which one might 'pierce through' to an actual object of signification, the ultimately spiritual character of being.

Drawing on contradictory characterizations of Carlyle by Emerson and Nietzsche, J. Hillis Miller has suggested that Carlyle's oblique and occluded language can be attributed 'to his situation as a human being, to his strategy as a writer, and to the meaning of the doctrine he preached'.[36] In terms of the doctrine, Miller points to a truth of the doctrine that cannot be written, but which nevertheless is approached through writing:

> If for Carlyle, the highest cannot be spoken of in words, and if the aim of *Sartor Resartus*, which is precisely words, words on the page to be read, and by no means simply gestures, is to speak of the highest, which clearly *is* its aim, then that speaking must necessarily be of the most oblique and roundabout sort. It must be a speaking which, one way or another, discounts itself in its act of being proffered.[37]

That is, the indirection and endless deferral of metaphor and symbol in *Sartor* can be attributed to the object of its discourse, which, as an ultimate truth, necessarily resists signification. Thus the text represents its own failure of disclosure. Likewise, like Teufelsdröckh himself, the text must circle but can never definitively attach itself to its referents. Drawing on Miller, Tom Toremas concludes that '*Sartor Resartus*, the account of the translation of a fictional Clothes Philosophy, ultimately comes to a halt in a language that refers to nothing but the text it is intended to transmit'. Toremas takes this referential circularity as significant for the political critique of Romanticism that involves the aesthetics of the *Bildung* in a 'totalitarian temptation', which the text is thereby resisting in its circumlocution. By resisting the aesthetic of the Romantic *Bildung*, *Sartor* has resisted the impulse to totalitarianism that is intrinsic to the *Bildung*.[38]

While these readings do well to characterize *Sartor*'s style and rhetorical mode and to connect them with plausible narrative purposes, they do so by eliding openings in the text that call for the reader's reconstruction of the significance and purpose of figurative expression. In short, metaphor and symbol change their meanings and functions over the course of the narrative of *Sartor*, and these changes hinge on the stages of the protagonist's progress through a spiritual pilgrimage. While necessary for describing the obscure and indirect character of the Clothes Philosophy, for representing that which cannot be directly apprehended, the 'Hieroglyphical'[39] character of representation is also associated with Teufelsdröckh's state of 'enchantment', with Teufelsdröckh's deeply roiled and melancholic dark night of the soul, the period of the 'Everlasting No'. In 'The Everlasting Yea', we later learn that Teufelsdröckh's long wandering in unbelief has indeed been a state of enchantment; he was 'entangled in the enchanted forests, demon-peopled, doleful of sight and of sound ...'.[40] In the 'Symbols' chapter, as well, lest we ascribe this reading to the Editor alone, Teufelsdröckh describes himself in this period as having been 'purblinded by enchantment'.[41] Furthermore, in 'Natural Supernaturalism', enchantment is linked in the past tense with what were its metaphoric and symbolic snares: 'Phantasms enough he has had to struggle with; Cloth-webs and Cob-webs, of Imperial Mantles, Superannuated Symbols, and what not'.[42] Thus, enchantment, with its entire phantasmagoria, characterizes a condition that calls for *Sartor*'s hyper-figurative language, and ultimately, its failure of representation.[43] Interestingly, in what amounts to an inversion of the standard secularization thesis, rather than the period of faith prior to its loss, Carlyle has figured the *loss* of faith as the state of enchantment.

Teufelsdröckh's spiritual journey also necessarily involves disenchantment, however, which changes the character of representation. As we learn in the 'Everlasting Yea', 'The Centre of Indifference' represents this disenchantment: 'Here, then, as I lay in that CENTRE OF INDIFFERENCE; cast, doubtless by benignant upper Influence, into a healing sleep, the heavy dreams rolled gradually away, and I awoke to a new Heaven and a new Earth'.[44] At this moment, Teufelsdröckh realizes that the enchanted forests have been cleared. The Editor figures it as the casting out of the demons of materialist philosophy: 'We should rather say that Legion, or the Satanic School, was now pretty well extirpated and cast out, but next to nothing introduced in its room'.[45]

If *Sartor Resartus* were an allegory corresponding to the standard secularization thesis, that is, if it were a straightforward narrative of disenchantment, progressive religious decline, and the eventual elimination of belief, and nothing else, the novel would end at this point. We have a newly disenchanted, secular subject (one, albeit, who has gone through an extra step where enchantment involved adherence to a materialist philosophy rather than immersion in a Christian cosmology). He is now able to identify and empathize with his fellow human beings, without recourse to divine injunction. He subscribes to a morality that does not depend on transcendence. Yet, as Joshua Landy and Michael Saler suggest, disenchantment is often accompanied by or may produce re-enchantment.[46] In *Sartor*, re-enchantment does occur – by way of 'Natural Supernaturalism', which, at least according to the Editor, appears for Teufelsdröckh to have opened 'to his rapt vision the interior celestial Holy of Holies [which finally] lies disclosed'.[47] Here at last the pilgrim reaches a spiritual destination, which inverts the teleology of standard secularization with the added dimension of the desecularization of the secular, or re-enchantment. Despite the Cloth-webs and Cob-webs, despite the winding and wending of metaphoric and symbolic excesses, 'yet still did he courageously pierce through'.[48] Given a new framework of vision, the world discloses its substantive spiritual character. Miracles are no longer esoteric and occluded, as they had been under confessional Christianity; at least to Teufelsdröckh, they are immediately apprehensible. They are no longer understood as violating the (albeit only partially understood) laws of nature. One no longer requires the supernaturalism of Christianity in order to believe the miraculous; miracles are a matter of seeing differently, seeing with new eyes. The world as it stands is miraculous. The supernatural is revealed immanently, within nature. At the same time, however, nature has been transfigured; it has attributes of the infinite, a character formerly reserved for deity: 'Nature remains of quite infinite depth, of quite infinite expansion'.[49] The divine is visible: 'the Divine Essence is to be revealed in the Flesh'.[50] Rather the concealing the spirit, that is, the flesh now discloses it. The symbols of nature are innumerable, and the Prophet is happy to be able to read a few lines, while understanding them through science becomes a matter of interpreting a few 'Recipes' of a 'Volume' that is more than a 'huge, well-nigh inexhaustible Domestic-Cookery Book'.[51] The implication is that for the man of science, nature is nearly incomprehensible in its vastness and his formulas barely scratch the surface; but for the seer, the whole is immediately apprehended; first, one must have learned to defy 'Custom', or the old religious clothes, which the dark night of the soul allowed the overcoming of; but one must also have learned to see anew, with fresh spiritual vision. Naming has now become a matter of obscuration; names 'are but one kind of such custom-woven, wonder-hiding Garments'.[52] This includes a naming of the divine, now accessible to the vision of the seer. Previously the only means of representation of that which remained unseen, language now functions in part as a 'Clothes-Screen'[53] for that which now can be seen without its aid.

Natural supernaturalism, that is, represents revelation within the secular, the disclosure of an immanent divinity in the natural; it is a re-enchantment of the secular mode made possible by the foregoing enchantment-disenchantment. *Sartor Resartus* is an enchantment-disenchantment-*re*-enchantment narrative made possible under a new secularity.

Carlile: the secular as scientific materialism

Like Charles Bradlaugh after him (whose apprenticeship in freethought took place partly under the tutelage of Carlile's widow),[54] Richard Carlile evinced an active belief in what we can now descry as a variation on the standard secularization thesis.[55] Carlile's narrative of secularization and the value it places on the secular *qua* secular, so at odds with Carlyle's views on secularization and the secular, may be summarized as follows (the irony of which is only slightly exaggerated here): A providential instrument of Enlightenment rationality, the printing press had arrived 'like a true [secular] Messiah' to liberate the mass of humanity from the 'double yoke' of 'Kingcraft and Priestcraft'.[56] Given this secular *deus ex machina*, the inevitable progress and diffusion of scientific knowledge must follow, heralding the ultimate dissolution of superstition. As such, Christian religious belief was soon to be extirpated from the public and private spheres in a millennial secularization procession, amid resounding hosannas and hallelujahs. Notably, for Carlile, the secular is figured in religious terms.

Richard Carlile was a publisher of cheap periodicals, including the *Republican* and the *Deist* (both founded 1819); he also published books, such as Thomas Paine's *Age of Reason*, and Elihu Palmer's *Principles of Nature*.[57] He contributed to 'the coalescence between revolutionary politics and infidelity' that became characteristic of a segment of radicalism from the 1820s.[58] His 'neo-Malthusianism', treated briefly in Chapter 3, made him one of the first authors and publishers to openly advocate birth control and women's reproductive rights in an English periodical and book.[59] Carlile drew his scientific notions from Baron d'Holbach, Palmer, and from William Lawrence, the professor of anatomy and physiology at the Royal College of Surgeons, and others. Their materialist explanations served his anti-clericism, his calls for the abolition of official state religion, tithes, and sinecures of all kinds. In *The Republican,* Carlile echoed Lawrence's materialist explanations of mind – 'the mind does not appear to be any thing distant from matter composing the brain' – and extended his argument to the replacement of religion by science – 'until you hear natural and experimental philosophy, or the works of Helvetius, Mirabaud, Paine, &c. delivered in the churches, you cannot hope to derive much benefit from Reason and Science'.[60] After publishing *The Age of Reason* and Palmer's *Principles of Nature,* the attorney general, given evidence supplied by the Society for the Suppression of Vice, prosecuted Carlile for blasphemous and seditious libel. For his account of the Peterloo Massacre in *The Republican,* he landed in prison for blasphemous libel and was fined £1,500. While in Dorchester Jail, he edited and produced most of the prose for *The Republican,* and composed his well-circulated *Address to Men of Science* (1821).[61]

As James Epstein argues in *Radical Expression,* Carlile 'offered a possible point of convergence between plebeian rationalism and radicalism and the concerns of certain scientists'.[62] In particular, although a fierce critic of contemporary and historical 'Men of Science' and contemporary science in general, he became a champion of William Lawrence, after the latter delivered a series of lectures attacking John Hunt's vitalist theory of life in 1817.[63] This foray inaugurated the early nineteenth-century vitalism-materialism debate, which focused on the question of life: was life a substance or vital influence imparted on matter from without (vitalism), or was it autotelic by virtue of an auspicious set of material conditions, including organization (materialism)? Vitalism accorded well with Christian theology in which God imparted life to lifeless matter, whereas materialism was associated with godlessness. Arguing the case for what was rightly taken for a materialist conception of life, Lawrence drew the condemnation of the medical community, and was attacked directly by John Abernethy, senior professor of anatomy and physiology at the Royal College of Surgeons, among others.[64] Lawrence was forced to withdraw his published *Lectures* on penalty of being stripped of his gowns. Exploiting the law forbidding the copyrighting of banned material, Carlile and others soon republished Lawrence's *Lectures,* the legal ban thus ironically making them more widely available.[65]

In light of the English theocratic state that Carlile identified as the impediment to an egalitarian social order, the role that science played for his radical politics is comprehensible. Steven Shapin and Arnold Thackray explain the convergence of science and such socio-political interests in connection with Manchester artisans. Their assessment of the social basis of science in Manchester and its appeal to the marginalized can be profitably applied to Carlile:

> The adoption of science as a mode of cultural self-expression also depends on a particular affinity between progressivist, rationalist images of scientific knowledge and the alternative value system espoused by a group peripheral to English society ... [an] alliance between science, peripheral status, and progressivist philosophy.[66]

Carlile's peripheral status and the charges of blasphemy that mounted against his publishing ventures may explain his adherence to a progressive version of scientific knowledge. Progress in science signaled the overthrow of the conditions that led to his harassment and arrest, as well as promising to bring about conditions most favorable to his social milieu as he saw it. Yet, as James Epstein notes, Carlile was remarkable not so much for his 'commitment to Enlightenment science', *per se*, but specifically for 'the consistency and rigor with which [he] came to embrace scientific materialism and to reject all expressions of religious belief'.[67] Under persecution for blasphemy, and excluded from the broader public sphere, Carlile figured his exclusion in the most extreme terms that he might imagine – in terms of a materialism that had been associated with revolutionary French politics, and was thus condemned, but which now represented an internal threat to order.[68] This was a 'new materialism', based on 'a picture of the universe dissolving and recreating in a ceaseless flow of material processes', as opposed to the Newtonian world picture of a splendid machine.[69] This sense of materialism as involving a malleable, shape-shifting matter was central to Carlile's politics; changing social and political conditions accorded well with and was predicated upon a conception of the natural order as subject to transformation.

As Jon Klancher has remarked, Carlile may be considered a participant in the early nineteenth-century version of the late twentieth-century 'Science Wars'.[70] In the context of the late twentieth century disputes, Carlile's epistemology would likely have been derided and dismissed as 'naïve empiricism' – for failing to account for the social and cultural mediation of scientific knowledge and for supporting a dominant positivism – science seen as detached from and irresponsible to its social context. But his epistemology had a different status and posed different problems for nineteenth-century opponents of a naturalism unhinged from its theological foundations. In his *Philosophy of the Inductive Sciences* (1840), William Whewell would later counter such views, protecting science from such thoroughly sensationalist, materialist conceptions as Carlile's. Whewell's epistemology, under which all observation was necessarily 'theory-laden',[71] came not as a counter to a dominant positivism, however, but rather as a defense against an emergent alternative to the received framework – radical materialism. Arcane knowledge (posited variously as 'innate ideas' or 'intuition') was protected by a chosen few who would then impose their authority upon the majority.[72] A strict empiricist methodology with materialist underpinnings threatened to undermine the authoritarian foundations of natural theology and the state religion that held sway over nearly every major educational institution in Great Britain.

In *An Address to the Men of Science*, Carlile advanced a materialist, secularist science wielded as a counter to the 'Kingcraft and Priestcraft' of the theocratic state, with its sinecures derived from taxes and tithes, and protected by seditious and blasphemous libel law.[73] *An Address* invoked an unwavering recourse to a materialist ontology tied to an empiricist epistemology within which observation confirmed the 'natural' basis of human equality and an egalitarian access to knowledge and the truth. Science was a system of observation the results of which should be 'open to all', and any claims made in the absence of observation based in nature, such as the cosmogonies of the clergy, were 'the work of ignorant imposters' and equivalent to 'madness'.[74] Carlile called on 'Men of Science to stand forward and unfold their minds upon this important subject [religion and its hold on the state]'[75] and to expose to the 'truth' derived from scientific observation the whole corrupt system of laws and institutions based upon 'superstition' and 'mythology'. Science was the key to the destruction of what Marxists refer to as ideology, which kept the people in thrall, and maintained the conditions of oppression:

> It is reserved for the Man of Science to rid mankind of this horrid ignorance and credulity, and to impress upon their minds the all-important subject of scientific knowledge ... All tyranny, oppression and delusion, have been founded upon the ignorance and credulity of mankind. Knowledge, scientific knowledge, is the power that must be opposed to those evils, and be made to destroy them. Come forward, ye Men of Science ... grasp at tyranny, at oppression,

> at delusion, at ignorance, at credulity, and you shall find yourselves sufficiently powerful to destroy the whole, and emancipate both the mind and the body of man from the slavery of his joint oppressors.[76]

The state was so entrenched in the religious ideology that 'brutalizes' society 'by setting its members one against another, upon different points of belief, all of which are proved to be erroneous and to have no foundation in Nature' that there could be 'no separating the one [religion] without revolutionizing the other [the state]'.[77] Thus, the secularization of the state would amount to an economic, social, and political revolution. And science was the means by which such a revolution would be eventuated.

For Carlile, the young William Lawrence represented the sole exemplar in his age of the kind of courage necessary for scientists to overcome religious ideology and assert the truth discovered by the 'second scientific revolution'.[78] Others had lacked the conviction of their beliefs and had left Lawrence defenseless and alone:

> Yet when that spirited young man, Mr. Lawrence, having obtained a professor's gown in the College of Surgeons, shew a disposition in his public lectures to discountenance and attack those established impostures and superstitions of Priestcraft, the whole profession displayed that same cowardly and dastardly conduct, which has stamped with infamy the present generation of Neopolitans, and suffered the professor's gown to be stripped from this ornament of his profession and his country, and every employment to be taken from him, without even a public remonstrance, or scarcely an audible murmur![79]

Other scientists had made discoveries in their fields that gave the lie to the theological conceptions of the universe. Given what they knew as the 'truth' about nature, it was incumbent on the men of science to band together, and under the cover of Carlile's protection if necessary, to assert the knowledge that clearly contradicted the theology of the clergy. Yet, except for Lawrence, the prominent men of science had remained silent on the anti-theological implications of science.

For Carlile, the great revolution in chemistry involved its thoroughgoing reductionism, the very reductionism that Carlyle derided and found so wanting for understanding nature. 'Every species of matter has been brought to dissolution, and its elementary properties investigated, by their crucibles and fires, or their galvanic batteries'.[80] Quite unlike

Carlyle, Carlile celebrated this reductionism as a great accomplishment. He lauded the then-President of the Royal Society, the social and intellectual climber, Humphry Davy, as the foremost exponent of this new chemistry, the science most important for demonstrating materialist principles. As one of the first men of science in England to acquire an independent income based on research,[81] Davy was the perfect candidate to 'stand forth'.

Astronomy, too, undermined the cosmogonies of the theologian. With characteristic literal-mindedness, Carlile suggested that since astronomers had not located heaven and hell with their telescopes, and because matter was deemed the only reality, such places must not exist:

> Now, from the present state of astronomical knowledge, and from the deep research that has been made into Nature and her laws, we have a moral and demonstrable conviction, that all cosmogonies are but the idle fictions of the human brain, and all the tales about heaven and hell as definite places, are from the same source.[82]

Given its demystification of deep space, Carlile supposed that astronomy had destroyed the credibility of religious doctrine. Yet, again, astronomers had not made manifest the implications of their discoveries.

The second object of Carlile's *Address* was to revolutionize education on the basis of a materialist epistemology. With a facetious and droll irony, Carlile called for the 'conversion of priests to professors in the various departments of this science [chemistry]'.[83] The study of classical literature and mythology, and the drilling of moral doctrines based on theology, would be replaced by the study of chemistry, astronomy, geography, geometry, and mathematics. Even the study of language would be undertaken through the 'medium' of science.[84] Carlile's radical science education marked the limit of revolutionary pedagogy among the '[m]any new plans and schemes for education [that were] daily starting up' during the period.[85] But even relatively moderate advocates for universal or near universal education, such as Henry Brougham, were often associated, usually by detractors, with such revolutionary goals, and opposed on these grounds. In his pamphlet denouncing Brougham's *Practical Observations*, for example, 'A Country Gentleman' pointed to the French Revolution as the result of the kind of plans that Brougham had in mind. He ended his tract with a declamation that 'that amiable gentleman, Mr. Carlile, is Mr. Brougham's very honorable associate, and very meritorious co-operator, in this "beautiful system of" – Philosophy!!!' The point was delivered as if the final blow, settling once and for all the controversy regarding the education of the working classes. If Brougham sanctioned the likes of the materialist and atheist, Carlile, his plans could be nothing less than pernicious to the foundations of the state and civil society, and should be vigorously opposed.[86]

As a possibility within an emergent secularity, the significance of Carlile's materialism was in fact its radical alterity, the sense in which materialism represented a radical other, which was now candidly and boldly expressed from a distinctly plebeian standpoint against a background of establishment reaction and persecution. This radical alterity – a rhetorical space for which had been opened partly by the atheism debates of the previous decades and partly by Thomas Paine's deism – was adopted and adapted by Carlile to suit his particular circumstances and purposes.[87] The emergence of a 'counter-public' in which Carlile's uncompromising materialism could be made accessible to a plebeian readership, ultimately and ironically was contingent upon the preponderance of theological opposition, and thus took expression in and against religious language.[88] This opposition was presented not only in state law and state-Church doctrine, but as importantly, in the hegemonic public sphere, a sphere from which Carlile was largely excluded. Scientific materialism, long considered the philosophical source of French revolutionary politics, as fueled by the *philosophes*, was precisely 'radical' in the sense given the term by modern chemistry; it signaled the instability and apocalyptic character that Carlile wanted his discourse to represent. Carlile embraced scientific materialism specifically for its radical valences and threatening oppositional potentiality. It suited his purposes for menacing his opponents with the representation of a shaken and revolutionized social order. (Subsequent chapters in this book, however, make clear that scientific materialism carries no *necessary* or inherent political or social meaning. It could be harnessed to the objectives of social stability and a gradualist, middle-class 'meliorism', or even, as in the case of social Darwinism, to reaction.)

Related to the issue of materialism's meaning is the question of how and why Carlile could understand science as inevitability secularizing – how, that is, he saw the second scientific revolution as necessarily impressing secularization upon the broader social order. And what was the obverse: why did he view celebrated scientific achievements as contingent upon secular premises? That is, why was science presumed secular? These questions may be answered by pointing to the expectations for secularism in connection with science as against religion. Arguably, Carlile inherited from the Enlightenment what we now refer to as the 'conflict thesis' of science and religion, an understanding that involves the pair locked in battle. Under this thesis, advances in 'science' are seen as threatening religious conviction and thus prompting a backlash and an inevitable and progressive retreat of 'religion' from the domains of knowledge and even the public sphere. As historians of science have suggested, this model often involves caricature, relying not on even-handed treatments of specific cases, but rather on reductionist and motivated projections of positivist ideals onto historical and contemporary interactions, while assuming positivist teleology.[89] Carlile's version of the science-religion relationship was based upon a wishful scientistic proto-positivism aimed at *promoting* science as secular (and secularizing), and perhaps more importantly, the secular as inherently scientific. Science represented a potentially authoritative weapon to be wielded in the battle for authority in the public and political spheres as against religion, as part of the wider battle over ideological, political and material resources. At the same time, a secular science represented the essential condition under which such goods would be forthcoming. Science was deemed to be secular, and the secular would gain from science's cultural authority. In some sense, for all their differences in social and professional status as well as rhetoric and tone, and despite the fact that he was not a 'man of science' himself, Carlile's mobilization of scientific materialism anticipated the uses Thomas Huxley, John Tyndall and others would make of 'scientific naturalism' for cultural authority and professional status later in the century. (See Chapter 4.[90]) Insisting that the professional scientist necessarily demonstrate a commitment to scientific naturalism, they made scientific naturalism a shibboleth for the wielding of scientific authority. The secular was thus a wedge for dividing legitimate science practitioners from imposters. In the early 1820s, however, the scientific naturalism that emanated from the radical press of Richard Carlile was a coarse, brazen materialism, and, perhaps most importantly, an epistemological and ontological position unaffiliated with the emerging professionalization of science in Britain. Carlile's deployment of scientific materialism should be understood in terms of his position as relative outsider, and in terms of the dominance of theological convictions in the Indian summer of natural theology.[91]

In fact, Carlile's faith in the men of science was largely misplaced at this historical moment. With respect to Humphry Davy, as David Knight has shown, given his earlier researches in Agriculture at the new Royal Institution and his marriage into the aristocracy, by the early 1800s, Davy had already become congenial with the landed gentry. Well before 1821, he had abandoned his radical republicanism as well as his associates in Bristol and the Tepidarian Society of London.[92]

Quite contrary to Carlile's leveling objectives, 'Davy's vision of society involved technical progress yielding increased wealth and leisure, with a continuing unequal division of property'.[93] More to the point, Davy's elemental chemistry had not led him to assume a materialist ontology. His researches into the electrical properties of elements and compounds had instead brought him to an opposite conclusion:

> Chemical properties clearly did not depend in any simple way upon material components. Davy took pleasure in this, because it was important for his generation to establish that science and political revolution were *not* linked. The ideology behind the French Revolution was supposed to be scientific: Diderot, d'Alembert and Voltaire had all seen science as a weapon to use against the *ancien regime*. Science, materialism and the Reign of Terror were all connected in the English mind, and to show that chemistry depended on still-mysterious forces or powers [electrical properties], rather than matter, not only went better with Romantic beliefs about how the world worked, but also with patriotism.[94]

As the case of Davy makes clear, the findings of the new chemistry need not to have led inevitably to materialism, and science should not be understood as necessarily demonstrating and propagating the secularizing tendencies that Carlile saw in it. The connection between science and the secular was (and remains) conditional rather than 'natural' or inevitable. In particular, had he been better acquainted with the great chemist – his aristocratic worldview and his philosophical conceptions drawn from chemistry – perhaps Carlile would have understood that his hopes for having Davy stand forth as an exponent of his demystifying scientific materialism would not likely be realized. Or perhaps he knew this. However, in *An Address*, Carlile confessed that he had never attended a scientific lecture or witnessed a scientific demonstration like those orchestrated by Davy at the Royal Institution, for which the latter had become famous; rather, he had relied strictly on self-directed reading to glean his knowledge and to draw his own conclusions about science.[95] From the first page of his *Address*, that is, Carlile was clearly marked as an outsider. In fact, his scientific materialism drew its social meanings and political valences from such exclusion.

In terms of astronomy, the gentlemen of science, who were generally better placed to make astronomical observations and to pronounce upon them with authority, addressed such objections specifically, arguing for natural philosophy's awe- and belief-inspiring effects. In *A Preliminary*

Discourse on the Study of Natural Philosophy (1830), John Herschel, philosopher of science and son of famed astronomer William Herschel, appeared to have astronomy in mind when he argued against:

> the objection which has been taken ... against the study of natural philosophy, and indeed against all science, – that it fosters in its cultivators an undue and overweening self-conceit, leads them to doubt the immortality of the soul, and to scoff at revealed religion. Its natural effect, we may confidently assert, on every well constituted mind is and must be the direct contrary ... it places the existence and principal attributes of a Deity on such grounds as to render doubt absurd and atheism ridiculous.[96]

Herschel was not at all original in staking such counter-claims for science against its theological opponents and radical proponents, both of whom often associated science with skepticism and materialism. Defenses against the accusations of skepticism and radical materialism had been well established by the time of the *Address*. Decades before, William Paley had decidedly entered the atheism debates with his *Natural Theology* (1802), commending natural philosophy for its confirmation of the deity, and espousing natural theology as the only defensible framework for the study of nature.[97]

As such, with the support of natural theology and a nascent philosophy of science, the 'men of science' took precautions against such conclusions being drawn from scientific studies as suggested by Carlile. The dominant guarantors of epistemic credibility in the hegemonic public sphere had largely preempted Carlile's declamations against theology, likewise rendering claims on science for scientific materialism and secularizing objectives rather impotent. His claims for the autonomy and secularizing tendencies of science were intimately linked with his own isolation from a hegemonic public sphere, a sphere he sought to enter, but which barred him primarily due to the very views that he espoused. However, as we shall see in the following chapters, the 'secular' would not always be so unsuccessful. But even in his own right, Carlile was successful in promoting secular ideals to working-class readers in the Zetetic movement, to which I now turn.

Science and secularism for the working classes

As Joel H. Wiener has pointed out, the 'republican and infidel' followers of Carlile's writing and publishing came mostly from amongst artisans and other workingmen and women in such towns as Manchester, Halifax, Huddersfield, Bradford, Spitalfields, Shadwell, Leeds, as well as London and elsewhere. Carlile's infidel and republican movement was marked by shared interest in 'free discussion' and 'science', and by a general disdain for organized religion. Groups of 'infidels', 'deists', 'atheists', and 'materialists' formed reading groups and societies with such grandiose names as 'the Society for the Promotion of Truth', the 'Friends of Rational Liberty', and the 'Zetetic' societies.[98]

Of these ideologically diverse groups, the Zetetic societies were the most significant in size and geographical reach. Mobilizing the Greek term meaning 'to seek for', with its implied meaning of skepticism, the Zetetics were 'quintessential infidel groups' devoted to 'seeking the truth' and pursuing an 'analytic mode of argument and demonstration' in Sunday evening discussions and lectures on science, philosophy, and theology.[99] The Zetetics embodied Carlile's ideals, as expressed in his *Address*, of replacing religious services with scientific discussions, lectures, and demonstrations. Zetetic societies were established in Edinburgh, Glasgow, Aston-under-Lyne, Salford, Stalybridge, and London. The Zetetic societies were forerunners as well as more ideologically radical versions of the educational institutions for workingmen represented by the better organized and better funded Mechanics' Institutes, which began with the Glasgow Mechanics' Institution in 1823.

For Carlile, the *results* of science were 'open for all'. Yet the intellectualism of the Zetetic movement came progressively to divorce itself from the populist bearings of the radical movement. Carlile foisted heroes of the intellect before his readers. In an obituary, he trumpeted the former carter Richard Hassell as an example of an artisan who had, by intellectual rigor and achievement, proven himself 'one of those village geniuses' who had risen above the 'clod-like brains' of his fellow artisans.[100] Hassell had contributed as a correspondent to the *Mechanic's Magazine* and co-edited the *Newgate Monthly Magazine* before his early death. Carlile touted Hassell as a worker who had escaped manual labor and become an intellectual hero. Indeed, minus the verse, Hassell appears as a kind of plebeian Shelley among the Zetetics.

Carlile's followers constituted a minority contingent within radicalism.[101] Their atheist and materialist zeal did not accord with much of popular radicalism's dependence on constitutionalism and arguments supported by Christian morality. Hunt and Cobbett, for example, distanced themselves from Carlile's atheism. Throughout the twenties, the rationalist republicans also diverged greatly from many of the forms of popular expression that had served radicalism since the 1790s. Figurative language, public theatrical display, 'the tavern world of profane sing-songs and outrageous parodies', were rejected in favor of a scientific literalism, a 'correspondence' notion of language,[102] the likes of which would be developed later by the mid-twentieth-century logical positivists. As the decade advanced, the Zetetic movement and Carlile's *Republican* came to emphasize philosophy and epistemological polemics over popular politics. In general, 'terms such as "scholar", "genius", and "philosopher" carried particular weight within Zetetic circles'.[103] The emphasis on ideological warfare over populist or even representative politics that would characterize the later secular movement was in formation in the Zetetic circles of Carlile and company.[104] This emphasis also narrowed the rationalist segment of radicalism and made any kind of simple notion of class solidarity impossible to maintain. That is, the Zetetics formed a public that did not fit comfortably within any developing sense of class. While acknowledging their backgrounds as artisans and mechanics, the Zetetics attempted to enter the world of letters as intellectual peers of those they sought to oust, seeking to compete with those in established intellectual circles that they detested in order to prove their mettle. As Epstein observes, 'Zetetics were frequently caught between cultural worlds, stubbornly seeking recognition from the very culture they sought to dislodge'.[105]

Scientific heroics would not be accomplished by the ordinary observations of the day worker, but rather only by those, of whatever class, who had mastered knowledge through rigorous study. Scientific knowledge was seen as a 'privileged knowledge' to which the plebeian radicals craved access.[106] In *An Address*, Carlile had extolled the scientific worth of such men as Davy, who was not an aristocrat, but ever 'loved a lord', and Lawrence, who was an aristocrat but had elsewhere claimed science for the artisan and mechanic as well. While science was 'the peculiar province of Mechanics', the realm of science was after all posited as class and value-free. But entrance for the artisan or mechanic was predicated on overcoming the obstacles posed by a life of labor and the adversity occasioned by poverty. The realm of the intellect did not overlap with working life, although it could originate there. Knowledge was a route to independence from the life of servitude among members of the working classes, but was not a guarantee of escape from its conditions. Education, which for Carlile could only mean scientific education, was the means to overcome the *intellectual* servitude suffered by the working classes. Carlile sometimes failed to mention that the culture of print had forged his celebrity status in radical circles; he claimed that education itself was the sole means of escape from manual labor:

> The progress of knowledge affords to every man the genius to educate himself, and it is by education alone that the majority can be brought out of a state of servitude to the minority... But as yesterday I was a journeyman mechanic, subject to many oppressions; today I feel, that I am the equal of, and independent of, every man in the world. Education alone has made this change; for, I am neither better clothed, nor better fed, than I was occasionally as a journeyman mechanic, and possess but little more of what is called property.[107]

Carlile's rhetoric was echoed closely by other educationalists, such as John Robertson, editor of the *Mechanic's Magazine*. Drawing on Bacon's famous equation, 'knowledge is power', the strains of the Carlilean scientific radicalism could be detected in the subsequent plebeian knowledge movement, as reformers and opponents knew only so well.[108] The Zetetics had not expressed the pride in labor that characterized later periodicals devoted to the artisans' and mechanics' knowledge movement. But while reformers softened or hid the edge of Zetetic intellectual elitism and jettisoned any indication of atheism, probably to quell the fears of the middle-class or gentry reformers enlisted for their aid, the theme of knowledge as a means of liberation and independence was retained and elaborated. Plebeian knowledge was promoted as a tool that, while claiming to be value-free and politically neutral, could be used for cultural and political authority and power.

Although the Zetetic movement tended towards elitism, its social superiors rightly saw the claims for cultural authority as having leveling objectives. After all, the Zetetics posed a challenge – albeit attenuated by an increasingly philosophical discourse – to established authority, involved as it was with the state Church. Both the proponents and opponents of plebeian scientific education anticipated the specter of radical science as propounded by Carlile, which itself was adumbrated by the residual fear of French revolutionary politics and its underlying materialist principles. Later plans by reformers for the scientific education of the working classes were framed so as to avoid the charges imputed to revolutionary science.

In helping to forge the new useful knowledge industry, most mechanics' periodicals and institutions, while distancing themselves from, or outwardly disavowing, the atheistic aspects of radical science, nevertheless adopted its egalitarian, inductivist ethos. Baconian science emphasized observation, and the ability to observe phenomena was not unique to gentlemen. Science, therefore, could not be their province exclusively. Further, observation revealed the real biological equality obtaining along the otherwise artificial social hierarchy. As the *Working Man's Friend and Family Instructor* later put it:

> Every person must have right or wrong thoughts, and there is no reason why a hedger and ditcher, or a Scavenger, should not have as correct opinions and knowledge as a prince or a nobleman. Working Men and working women have naturally the same minds or souls as lords, ladies, or queens ... [I]f anyone could have analysed or cut to pieces the soul of Lord Bacon, or Sir Isaac Newton, and that of a chimney sweeper, it would have been found that both were made of the same divine material.[109]

Much as Carlile had 'dissected' the opus of Bacon, taking what parts he liked while vituperating upon the rest, Bacon's aristocratic 'soul' is subjected to Bacon's own method of induction in an act of epistemological leveling. Observation is an instrument in the hands of a social surgeon, good for dissecting and 'cutting to pieces' the exclusivity of gentlemanly knowledge claims, as well as the pretenses of rank.

Beyond this *de facto* egalitarianism, promoters of plebeian knowledge advanced claims for differential access to experimentally derived knowledge based on the worker's standpoint within the means of production. The *Mechanic's Magazine* put forth special claims for knowledge deriving from mechanics and artisans whose practical experience afforded increased opportunities for unbiased observation, albeit lacking the necessary theoretical basis for drawing conclusions from them:

> Philosophers will frequently find reason to follow the advice of Bacon, who recommends them to avoid a hasty contempt of popular opinions. The mass of mankind will be found with few exceptions perfectly correct in points of observation, however erroneous the conclusions may be which they sometimes deduce from their premises. Men of science are too apt to treat as vulgar errors, facts which do not admit an obvious explanation, and thus neglect many interesting phenomena, observed by mankind at large, whose experiments ought to be the best, being directed by no favourite theories, and biased by no hypotheses.[110]

In contemporary terms, the observations of the working classes, so the claim went, were less likely to be biased by 'theory-ladenness', by a particular theoretical bias. Furthermore, such sources promoted a kind of standpoint epistemology well before it was elaborated in the work of György Lukács and appropriated by advocates for a feminist science or criticism of science.[111] The claim for the workers' unique standpoint for knowledge became another justification for the diffusion of theoretical knowledge to the working classes. Not only would the diffusion of knowledge 'improve' *them*, it would also advance science itself. Henry Brougham echoed the point in his arguments for the benefit to science of popular education:

> Indeed, those discoveries immediately connected with experiment and observation, are most likely to be made by men, whose lives being spent in the midst of mechanical operations, are at the same time instructed in the general principles upon which these depend, and trained betimes to the habits of speculation.[112]

Brougham folded together the benefits of educating the working classes in science with the interests of industry and the advancement of science itself.

On one hand, the arguments for the new knowledge industry were made on behalf of the working classes, for whom a natural right had been denied, and whom their masters and the government had kept in a state of ignorance. The working classes had also contributed by partaking in less than fruitful activities. On the other, science and society had ignored or discounted the knowledge already possessed by the working classes. The working classes were denied access to scientific theory and denied the recognition of their empirical knowledge. As John Robertson, the primary editor of the *Mechanic's Magazine* saw it, for this very reason, 'the Press, was called in to assist in the dissemination of a knowledge of principles among the working classes, and in obtaining from them, in return, those benefits which practice has it so much in its power to confer upon theory'.[113] That is, despite the dominant notion that the 'diffusion of knowledge' was unidirectional, promoters like Robertson (and Carlile) saw the knowledge industry in terms of a two-way diffusion or dialectic.

The significance of the Zetetic movement, despite (or because of) its humble, lower-class provenance and constitution, was that it promoted scientific knowledge as independent of and antithetical to religious conviction; it was an early attempt in Britain to fully secularize knowledge as such, and it saw science as naturally secular in its tendencies and implications. This was a form of mass secularism that would be taken up in the 1840s by Carlile's successors in the freethought movement (see Chapters 3 and 4), and would be continued by Carlile himself at the Rotunda on the south bank of London during the 1830s, where he lectured and Robert Taylor performed before fairly large audiences.[114]

Conclusion: secularity as plurality

As discussed above, Frank M. Turner has pointed to the increasing religious diversity of the nineteenth century, to the burgeoning new Christian sects and creeds, and the role that these may have played for the emergence of secular conversions in the period. These Christian sources of the secular paralleled other, more or less secular sources that also contributed to secularity broadly construed. This chapter has illustrated sources and characteristics of two heterodox positions, and, especially in the case of Carlile, underscored the social and political provenance and meanings of such heterodoxies. By focusing on the distinct heterodoxies of individuals occupying quite different social locations and driven by very different objectives, I have demonstrated a range of the *optative condition* that secularity made possible, or perhaps even made necessary – that is, a set of options faced regarding belief and unbelief, religion and the secular. These choices included, as we have seen, what we might now call 'individual mysticism',[115] as well as radical scientific materialism, among others.

While I have chosen to begin my exploration of secularity by discussing two (rather peculiar) individuals, and, in the case of Carlile, some of his followers, it should be clear that such belief pluralities help to delineate the wider structural forms from which they drew their meanings and vitality, and to which they contributed as sources. One of this book's premises is that pluralities, both religious and secular, depend upon and together constitute the optative condition of secularity. The pluralities that we note on the individual level are represented at the group and societal levels. David Nash suggests that a focus on the multiplicity of ideologies, rather than on the exceptionality of individuals, or even the peculiarity of groups (such as religious sects or denominations, secular societies, and so forth), should enable historians to overcome the dichotomy we might otherwise encounter between the study of individuals on the one hand and the study of groups on the other.[116] As we shall see, as opposed to appearing exceptional in the more or less recondite writings of Carlile and Carlyle, the secular may also be found within institutional frameworks, and met in scientific, social, and political movements.

In the next two chapters, I examine secularism in the contexts of gentlemanly geology and artisanal freethought. In neither of these spaces, however, should the secular be understood as the mere negation of the religious; rather, as Colin Jager points out, following Talal Asad, the secular and religious are always co-constituted; the category of the 'religious', in fact, is a product of the secular, against which the latter can be figured as 'neutral'.[117] As these chapters together will make clear, the content, social meanings, and political valences of the secular are by no means stable, nor is the secular ever neutral. The secular always emerges as a contingent, historically shaped and culturally mediated element and relation. Secularity exhibits a great deal of variation in the expression of the secular and its other, the religious, in the various domains where it appears.

2
Principles of Geology: A Secular Fissure in Scientific Knowledge

A study of nineteenth-century secularism would be grossly incomplete and nearly incomprehensible if it failed to address scientific knowledge production in the period. However, as Charles Taylor has suggested in *A Secular Age*, 'secularization', to the extent that it can be thought to have happened, cannot be understood as an inevitable effect of the 'rise of science', urbanization, bureaucratic rationalization, and so forth – that is, as a result of the processes of modernity itself.[1] Nor, with Owen Chadwick, can we regard the secular as a by-product of materialist science in particular.[2] For Taylor, the standard secularization thesis, among its many flaws, amounts to an instance of the *hysteron proteron* fallacy – a confusion of subject and predicate, of cause and effect – or of affirming the consequent: modern science (or urbanization, rationality, or what have you) is secular; thus, with the advent of modern science comes secularization. But this construction begs the question: how did science become secular before secularization, as such?

While the standard secularization thesis may mistake cause and effect, it also advances an untenable progress narrative, positing the secular as a 'natural' condition to which history has always been tending. Such a Whiggish narrative as the standard secularization thesis represents has been partially belied by history and rejected by historians,[3] while even losing favor among prominent sociologists.[4] However, although partially a product of positivist sociology and Whiggish historiography, the category of the secular nevertheless must be accounted for historically. Or as Laura Schwartz puts it in a related context, 'a study focusing on self-proclaimed Secularists actively engaged in constructing a secular public sphere, obviously does not allow for the category of the secular to be left unexamined'.[5] Similarly, treating the 'conflict thesis' of science and religion, Frank M. Turner has suggested that we should not discount the existence of real conflicts as among the reasons that such a historical model emerged in the first place.[6]

As I have been arguing, the secular is a necessary analytical and historical category, although it should not be reified or taken as inevitable.[7] The 'secular' and the 'religious' are mutually constitutive, context-dependent categories that exist as such only in relation to one another. Moreover, the conception of the condition of *secularity* encompasses both. The emergence of the secular may be understood in terms of an undergirding episteme, a framework within which the secular emerged and accrued cultural, intellectual and social authority, and was wielded for personal, professional and political purposes across various domains. At the same time, the secular has largely resulted in (or been the result of) the division and differentiation of various spheres of cultural authority. In knowledge fields, the secular amounted to a space developed from predominately religious frameworks, such as Natural Theology in nineteenth-century Britain. Further, the secular may be understood as a means of distinction within scientific discourse and practice, a tool for differentiation within fields that it altered, although not once and for all and certainly not according to the necessary progress narrative or *telos* suggested by Auguste Comte and his followers.

As this chapter aims to show, the emergence of the secular in science has been both contingent and context-dependent. In order to grasp the possibility, contingency, and significance of the emergence of the secular in geology, I undertake a study of the making of Charles Lyell's *Principles of Geology* (1830–1833), focusing on the publishing and periodical history leading to and including its publication, as well as coterminous publications from its publisher. Thomas Dixon has remarked that the context of publishing has often been a site for the wielding of knowledge as power.[8] This chapter shows how Charles Lyell's 'uniformity' – the principle that only forces acting in actual time at present could be admitted as explanations for past geological phenomena – was more than a principle of scientific observation; it was a demarcation device not only in geology but also in the fields of scientific endeavor and knowledge production more generally. The knowledge project conducted by Lyell in connection with the Murray publishing house and the *Quarterly Review* reveals an early (but not complete) fissure in the gentlemanly scientific consensus regarding just who would be authorized to produce scientific knowledge, and on what basis. From 1830 through the emergence of scientific naturalism at mid-century and beyond, this would be defined increasingly in terms of the secular. Uniformity can be understood as an early secular wedge wielded by Lyell in an attempt to divide the wheat from the chaff in the knowledge fields from the late 1820s on.

As I will argue, uniformity was also a normative policy. A reformist trope, uniformity worked as a form of secular social and political 'gradualism', a companion to the kind of naturalistic gradualism contemplated by Auguste Comte in France in the early 1830s, and a precursor to that promoted by George Holyoake with Secularism in Britain at mid-century. The Lyell-Murray knowledge project was undertaken in large part with the aim of confronting an existing plebeian public sphere – in which a proto-scientific rationality served the ends of a revolutionary project – and transforming it into an explicitly political mode of institutionalized knowledge production specifically tailored for the British upper classes, yet intended to inflect upon the meanings given to science by the promoters of radical and 'popular' science as well. As I will show, Lyell's rejection of then-extant materialist theories of evolution was motivated in part by the potentially radical consequences of the doctrine for the restive lower classes. Thus, the secular, far from being ideologically and politically 'neutral', as has often been claimed, was in fact overdetermined by cultural, ideological and political ideals, and framed within a *reformist* project designed to guard science from radical alternatives.

In the nineteenth century, the scientific fields did indeed undergo secular convulsions; geology was perhaps the first. Since Hutton's assertion in the late eighteenth century of deep time and endless (secular) cycles of the earth, and with the continual unearthing of the fossil series by an almost exclusively empiricist geological practice, nineteenth-century geology had braced itself against a fully systematized, naturalized epistemology. Yet, secularity would indeed erupt, and in fact may have demonstrated its most convulsive effects in the field that, paradoxically, concerned itself with the oldest of earthly phenomena.

In the first four chapters of Volume One of his *Principles of Geology*, Charles Lyell sketched a history of geology, tracing the development of the earth's study from ancient times to the early nineteenth century. According to Lyell, geology still labored under theological premises and explanations. In Lyell's account, geology might finally be freed from pre-scientific error with the eventual and inevitable discovery of the 'correct' theoretical principles, the very principles he set out to demonstrate in the work itself. But before geology could become a science, its practitioners had to recognize its 'legitimate objects'. For Lyell, the 'most common and serious source of confusion' had been 'the identification of its objects with those of Cosmogony', which was analogous to confusing history with 'speculations concerning the creation of man'.[9] Those studying the earth's surface had allowed for 'dramatic and even supernatural causes' – massive floods, earthquakes, a 'plastic force' in nature – and explained the otherwise inexplicable by reference to 'the origin of things'.[10] Lyell suggested that no such explanations could be allowed if geology was to become 'scientific'. Lyell's history of geology culminated with the founding of the Geological Society (1807), just before the contemporary moment.[11] But he made clear the association of his contemporaries with the unprincipled speculations of the past, relegating those not affiliated with his thoroughgoing theory of uniformity – those later dubbed 'Catastrophists' by William Whewell, the most important neologist of the time – to 'a history of absurdities'.[12] He associated his contemporary opponents with the errant tendencies of the past, while aligning his own views with the 'natural' tendency to arrive at scientific methods and principles. As Adrian Johns has noted, such positioning of one's work as the natural *telos* of history had been a long-standing trope in the making of science. Lyell performed a historiographical exercise in nineteenth-century geology similar to that of John Flamsteed in early eighteenth-century astrology, whose 'full history of astronomy from biblical times to the present, [was] designed to culminate in its own appearance'.[13]

As Stephen Jay Gould noted, the task of disentangling a history of geology from Lyell's own self-congratulatory account has been complicated by the fact that it had become the received or 'cardboard history', embedded as such over time as historians continued to reproduce it, almost verbatim.[14] More recent accounts have undertaken the process of restoring context to both Lyell and the figures he represented.[15] Historians of science have sought to 'examine Lyell in context of his place in the historical record, and recreate the economy of thought at the time of his own deposition'. Ironically labeling Lyell's history 'catastrophist', 'erroneous propaganda', and 'historical romance', scholars suggested that Lyell exaggerated the extent to which his *Principles* represented a clean break from history and its own contemporary moment.[16]

Nevertheless, Lyell's intervention, as well as his representation of it, is significant in its own right. To grasp just how, historians imply that its contexts of production and reception must be restored. Yet, despite its monumental status as a scientific tome, with the exception of passages in one major biography, and in scholarly introductions to the work, scant attention has been paid to the making of the treatise.[17] Such a study would seem to point to the answers that historians seek in contextualizing Lyell's work, and by extension, other such scientific 'revolutions'. More to the point here, however, such a study should show the contingent emergence of the secular, and demonstrate whether and to what extent it represented a 'catastrophe' in geology.

The making of *Principles* should not be considered as an individual effort, however. Recent studies in the history of science have underscored the importance of publishers, editors, printers, binders, reviewers, and others in the making of science. As Susan Sheets-Pyenson has metaphorically put it, publishers and printers 'acted as midwives in the creative process of bringing forth periodicals and books, [and] made decisions about which forms of scientific literature could survive in the market place'.[18]

Furthermore, Lyell's project must be seen in the wider context of a new popular 'knowledge industry' that burgeoned from the first two decades of the century.[19] The new knowledge industry to which Lyell's project responded drew on a tradition that 'encouraged participation by a wide range of people in the making of knowledge'.[20] Publishers, editors, critics, educational reformers, politicians, authors and others asserted their agendas within the new industry, positing competing claims for knowledge-making and dissemination. Sensing that nothing short of a scientific revolution was underway in Britain by the 1820s, Lyell and his publisher sought to shore up control of the sciences for gentlemen. Drawing in part on an earlier, unpublished version of this chapter for a chapter of his recent book, Jon Klancher remarks that 'Lyell's alternative, institution-based vision of scientific emergence was aimed to redirect the evolution of British science from that urban plebeian sphere to the provinces and the aristocrats' landed estates. Ultimately it would point to the emergent world system of the new scientific age'.[21]

The Lyell-Murray knowledge project advocated *naturalism* – the principle that all observable phenomena could be seen as entirely natural events – from the position of established opinion and within the most established institutions. Thus, the Lyell-Murray intervention represented a break within natural history and natural philosophy, with their reliance on the epistemological and philosophical framework of Natural Theology, inaugurating the differentiation and 'complexification' that Klancher discusses in his treatment of developing Romantic-age knowledge institutions, as well as the professionalization in the sciences to be undertaken by the science publicists and scientific naturalists later in the century. While deriving from elite culture, the project nevertheless called for and evidenced a divide within that culture.[22] The divide may be seen in terms of an early secular opening within gentlemanly knowledge production and publication, and moreover, as a preemptive tactic to ward off more radical alternatives arising from plebeian quarters.

The *Quarterly Review*: a scientific turn

The new, 'gentlemanly' knowledge project was initiated in the Tory *Quarterly Review* as the *Quarterly* took a scientific turn. John Murray, the *Quarterly* publisher, who would also publish Lyell's *Principles*, underwrote it. Lyell's science writing career began with articles on science and education that he contributed to the *Quarterly* beginning in 1825.[23] The editor, John Taylor Coleridge, the nephew of Samuel Taylor Coleridge, had invited him to contribute, and Lyell had immediately set to work on a review of Thomas Campbell's *Letter to Mr. Brougham on the Subject of a London University* (1825). This first article was conservative in orientation, as the *Quarterly* continued in its usual response to the educational reform questions of the day, and its ideological opponent, Henry Brougham at the *Edinburgh Review*.[24] Brougham and the *Edinburgh* campaigned heavily for a new, London University, ruthlessly attacking the exclusive, classical education at Oxford and Cambridge.[25] In response, Lyell defended the English universities, and by implication his own Oxford education, calling only for gradual modifications in light of the campaign for London. But he began his research for what would become a criticism of the English university and educational systems. While Lyell's educational politics were relatively conservative and remained so – true education remained the sole property of the aristocracy – in light of his own circumstances as a fledgling graduate trying to eke out a living, he began to consider university reform, and also, the newly emergent scientific and literary institutions cropping up across the country.

After graduating from Oxford, Lyell had moved to London to become a barrister. '[F]alling into the society of Lawyers, Geologists & other sinners',[26] he became a liberal Whig, advocated electoral reform, and probably believed in the disestablishment of the Anglican Church. James Secord has suggested that Lyell – coming as he did from a highbrow Tory family opposed to the Reform Bill – managed to find in science 'an indirect way of forwarding reform without betraying his father or his teachers [at Oxford]'.[27] Science and educational reform offered a 'safe' means by which he might enter into the spirit of the age, but more particularly, Lyell believed, it was one of the only means of efficacious action open to him. In 1826 and 1827, he published the first reformist articles on science and education to have appeared in the *Quarterly Review*.[28]

The first and arguably most important was 'Scientific Institutions'. Here Lyell heralded a second scientific revolution.[29] This early notice of a second scientific revolution connected the growth, specialization, and professionalism of natural knowledge to the status of empire.[30] The article was ostensibly a review of the transactions of the new scientific societies in Cambridge, Cornwall, Manchester, Liverpool, Bristol and Yorkshire. But Lyell used the occasion to deliver his own assessment of the state of science in 1826 England, and to offer his views for the reform of its knowledge enterprises. After the founding of the Royal Society in 1663, nearly a century lapsed before 'a national museum of Natural History [the British Museum] was founded in our metropolis' in 1759.[31] 'From the institution of the Royal Society in 1663, to the year 1788 ... no subdivision of scientific labour was attempted in our metropolis', until the founding of the Linnaean Society, which undertook the 'prosecution of the studies of zoology and botany in all their details'.[32] But soon after the founding of the Royal Institution in 1799, Lyell reported, numerous societies devoted to one or another division of natural philosophy had sprung into existence. The first decades of the century were marked by a veritable explosion in scientific activity and its specialized, metropolitan institutions: the College of Surgeons in 1800, the Horticultural Society in 1804, the London Institute for the Advancement of Literature and the Diffusion of Useful Knowledge in 1805, the Geological Society in 1807, and the Astronomical Society in 1821.[33]

But Lyell was careful to compensate for such metropolitan boosterism by noting the 'rise and progress of similar institutions in the provinces', which for him were even more important than the metropolitan. With this provincial emphasis, Lyell took the obligatory swipe at the *Edinburgh Review* and its recent campaigning for a London University based largely on supposed metropolitan superiority.[34] The new provincial societies included those founded in Manchester (1781), Cornwall (1814), Liverpool (1814), Cambridge (1819), Bristol, (1820), Yorkshire (1822), as well as 'many other institutions in our provinces, such as those of Newcastle, Bath, Leeds, and Exeter'.[35] Thousands thereby gained exposure to the natural sciences; but more importantly for Lyell, 'a new class of lecturers' had been born whose employments in the branches of natural knowledge had allowed them 'to enlarge and perfect their own knowledge'. Lyell saw the possibilities for a 'certain class of the community, to direct their minds and devote their lives *professionally* to these studies'. The provincial societies offered new theatres for 'native talent', and perhaps did more for knowledge production than their metropolitan counterparts.[36] Eventually, the rank of such societies would secure their place as objects of ambition even for men from the nobler classes. And, the implication was that every colonial possession, every new territory, was potentially a new theatre for talent and a new source for knowledge. Thus, the national intellect was tied to the growth and development of empire.[37]

By 1826, therefore, long before the 'science and culture' debates of Huxley and Arnold later in the century, or the putative 'science versus religion' disputes between the scientific naturalists and their theological opponents,[38] Lyell had already begun to delineate a model for the division of natural knowledge from cultural, including religious, enterprises, arguing that such a division was a prerequisite for the development of natural knowledge. Further, he outlined a system of professionals, lecturing to an enlarging public, and contributing original research to their fields of knowledge. This professional lecturer and researcher in natural sciences would be differentiated from the older model for natural philosopher. No longer an amateur, he would pursue scientific matters exclusively. Such a situation in science would benefit men like Lyell, who, although aristocratic, nevertheless struggled to earn an independent living. One would pursue natural science in an institutional setting in which both the production and exchange of knowledge would be accelerated. Additionally, the new breed of naturalist could effectively eschew religious affiliation, and eventually even the necessary conciliation between natural and revealed truth, as represented in the usual framework of natural philosophy, Natural Theology. Thus, the secular within scientific knowledge production was for Lyell tied to professionalization.

Lyell's correspondence shows that he was amused at having gotten a reformist article published in the Tory *Review*. 'I must not sport radical', he joked in a letter to his friend, Gideon Mandell, 'as I am become a Quarterly Reviewer. You will see my article just out on "Scientific Institutions," by which some of my friends here think I have carried the strong works of the enemy by storm'.[39] Lyell was being playful and triumphal, as he had already 'sported radical' within the *Quarterly*, into which he had carried 'the strong works of the enemy', referring undoubtedly to the kinds of educational and institutional reforms being proposed at the *Edinburgh* by the utilitarian and Whig, Henry Brougham. Lyell's gentlemanly status,[40] coupled with his skilled rhetorical maneuverings, had enabled him to smuggle a reformist article into a bastion of periodical conservatism. Likewise, it might have looked as if the *Quarterly* had finally conceded to, or in fact was proposing, some relatively 'radical' change.

He immediately set to work on his next project, an article about the Royal inquiry of Scottish universities, and sent a draft to William Buckland at Oxford and a friend at Cambridge. Lyell sensed the danger of treading on established opinion. He wrote to Murray about Buckland's treatment of his upcoming essay. 'Professor Buckland's censorship has been exercised as freely as by the new French Commission on the *Journal des Debats*. But barring the *unorthodox* parts he is pleased with it, & I shall take care under such correction not to frighten the nerves of the Q.R. readers which are only getting strength on these matters'.[41] We can see from this letter that Lyell and Murray had some sort of understanding about Lyell's objectives, or at least that Lyell assumed Murray to be sympathetic. A second letter seems to bear this out: 'My university Art. is at length finished but the sensitiveness of Ox. & Camb. is amusingly great & the softening down of passages where the naked truth came out too clearly, some more of which a letter from Oxford this morning made necessary, would amuse you if you saw my correspondence'.[42] By the publication of the first reformist article, Lyell was 'far on with the second, and hope[d] to get it out in less than three months'.[43] The *Quarterly* soon published his 'State of the Universities', which cut to the heart of English institutional establishments – the university system of Cambridge and Oxford.

'State of the Universities' represented another important first for the *Quarterly*, amplifying Lyell's earlier revision of scientific educational institutions, and extending it to the ancient university structure. Purportedly, the article was a review of the publications relating to a Royal Commission established to investigate the Scottish universities. Lyell used the occasion to undertake a historical and geographical survey of university education, and to evaluate English and Irish universities in particular. Not only did the Scottish universities resemble each other on basic assumptions, they also resembled the French, German, and Italian. Further, it was in the very points of agreement that they *all* differed from the English and the Irish. As such, the English and Irish systems exhibited peculiarities not seen in any of the others:

> There are three striking peculiarities in the system of education in England and Ireland without parallel in any of the other nations of modern Europe: First, the length of preliminary education, and the limited extent of the subjects it embraces: Secondly, the virtual exclusion of a regular course of study in the faculties of theology, law

> and medicine: Thirdly, the very incomplete subdivision of sciences among those on whom the whole burden of teaching is cast.[44]

Finding the English and Irish universities deficient in comparison with other systems, and indeed, in terms of their own history, he argued for the introduction of professional studies, and in particular, the increase in scientific studies, as well as their inclusion in the examinations. The upshot was the recommendation for new professors in the various departments of knowledge, especially in natural knowledge, thus enlarging the new natural knowledge class that Lyell envisioned.

As might be expected, what Lyell lamented most about the English system was the absence of studies in the natural sciences. While the taste for natural sciences is weak in most students, others have 'an irresistible, and as it were, instinctive propensity to cultivate such studies, and if no elementary knowledge be communicated in a scientific form, they will, nevertheless, follow the bent of their inclination; and what might, under a proper direction, have led to the improvement and exercise of the mental faculties, must often degenerate into a frivolous amusement'.[45] Such 'frivolous amusement' referred to the amateur status of the naturalist in England and implied the need for a dedicated class of science practitioners, who might be remunerated for their pursuits.

Lyell elaborated the unique benefits accruing to the study of natural philosophy. He first recurred to the usual arguments of Natural Theology. Given the conspicuous absence of such arguments from *Principles*, this may be surprising. In fact, his break with Natural Theology in *Principles* might be considered the initial fissure of this commonality within gentlemanly British science to date. But here he retained it to further his argument for studies in natural science. To their contribution to Natural Theology, he added that studies in natural science 'may, at least, be of high usefulness in future life, either for relaxation after intense study, or for refreshing and restoring the mind to a healthy state when suffering under disease of worldly disappointment'. Their 'slight connexion with human affairs' was in fact what rendered them particularly useful for relaxation and diversion, despite those who wielded this as 'a vulgar objection to their inutility'.[46] Lyell argued that the natural sciences – as opposed to history, politics, and poetry – were neutral with reference to human affairs and passions, and offered exercise of the mind upon matters regarded impartially and without prejudice:

> When we read history, we are presented with facts often distorted by political prejudices; and however distant may be the transactions

> from our own time, we are seldom indifferent and impartial arbitrators... The same remark applies to ethics and politics in general; they seldom afford a neutral ground, like the problems of physical science, where conflicting evidence may be tried fairly by its own strength, and the judgement formed by an habitual practice of examining proofs with an unbiased desire of discovering the truth.[47]

Lyell advanced an argument for the neutrality of science, an ideology of science as an activity that excluded passion, prejudice and bias, and especially partisan political fervor. This was not a new concept, and in fact had been suggested by Bacon. But here and now his argument for neutrality, for 'disinterested curiosity', also echoed that of other recent advocates for the advancement of science in education, such as Henry Brougham and the *Mechanic's Magazine*, and thus ironically took on a new political valence. In the *Quarterly*, where science had often been associated with French radicalism, the position of science as value-neutral was a challenge.[48] Lyell's point was to offer science to the aristocratic elite on terms that they could accept. Science should not be associated with politics. The importance of science to the growth of industry, moreover, made its study increasingly important.

Lyell's *Quarterly* articles of 1826–1827, taken together, outlined a model for the reform and modernization of knowledge enterprises within English educational and scientific institutions. First, Lyell called for the *separation of the spheres* – of the sciences from other cultural enterprises, that is, the fine arts and literature. The division of natural and cultural knowledge enterprises had to do partly with the requirement that institutions keep pace with the growth of knowledge, but the separation/differentiation of the spheres was also necessary in order to steer patronage to the sciences. Secondly, he called for a new *professionalism* in the sciences – a new class of lecturers, stationed both within the new provincial institutions and at Oxbridge. Third, he called for increased *specialization* or differentiation within the branches of natural knowledge. Specialization was required by the growth of knowledge, but it also yielded more professional positions within the emerging and older institutions. Given the *Quarterly's* former conservative resistance to almost all education reform, the model represented a significantly altered view of education for the publication and its readers.

Lyell's publications in the *Quarterly* groomed him as a popular science writer and reformer well before he ventured to write *Principles*. As he suggested to his sister Caroline, others at the *Quarterly* had pleaded his case. 'I find I have risen mightily in their opinion [Murray's and

Lockhart's] & Barrow,[49] who abuses almost everyone & whom I cannot bear, has assured absolute John [Murray] that I have shown more tact in rendering a scientific Art. popular & intelligible to the uninitiated than any writer he could find in town'.[50] Lyell's ongoing connections with the *Review*'s editors and its publisher allowed him to have his *Principles* printed on generous terms and at the sole risk of the publisher.[51] And although he claimed otherwise, such associations also allowed him influence over the selection of *Quarterly* reviewers for his first two volumes of *Principles*.[52] In exchange, the *Quarterly* paradoxically became a source of relatively forward-thinking science and educational criticism, almost overnight, and Murray gained a new author for promoting his version of gentlemanly knowledge in the new knowledge industry.

John Murray II and the Family Library

The grooming of Charles Lyell as science writer for the *Quarterly,* and the scientific turn of the periodical, paralleled John Murray's book publishing efforts within the new knowledge industry. The new knowledge industry involved numerous plans and campaigns for knowledge dissemination, including Mechanics Institutes, reading clubs, lending libraries, useful knowledge encyclopedias, useful knowledge periodicals, and most importantly for publishers, cheapened books. In 1826, the Society for the Diffusion of Useful Knowledge (SDUK) had been founded with the 'philosophical radical' and educational reformer Henry Brougham as chair. The SDUK published a Library of Useful Knowledge, Library of Entertaining Knowledge, and Library for the Young, the Working-Man's Companion, and the Farmer Series.[53] Almost from the outset, such middlebrow enterprises as the SDUK dominated the knowledge industry.

Given his outstanding reputation, John Murray had been the first choice of the SDUK for its Library of Entertaining Knowledge.[54] Murray agreed to publish the new series, and even contributed ten pounds himself. But he later withdrew his offer, possibly in response to Charles Knight's admonition that original publications intended for plebeian readers would not be self-supporting, let alone profitable: 'The millions were not ready to buy such books at a shilling, nor even at six-pence', Knight later wrote.[55] Undoubtedly, Murray also responded negatively to the 'universal education' objectives of the SDUK. The enterprise was turned over to Charles Knight, who printed the Library of Entertaining Knowledge and also the *Penny Magazine*.[56] Murray's series, the Family Library, owed its origin in part to the difficulties the

SDUK had encountered with the Library of Useful Knowledge. The 'inexpensive form was at odds with the expectation of genteel learning' and Murray hoped to fill the niche for cheapened books suitable for genteel readers.[57]

With the flood of cheap knowledge publications on the market from the 1820s, the 'plebeification' of knowledge, of which Coleridge had warned, was apparently coming to pass. John Murray feared that the control of knowledge might slip forever from gentlemanly hands. Unless a gentlemanly project intervened in the glutted market, new readers might take such publications as the SDUK's and others to be definitive accounts of knowledge in their respective fields. Thus, while SDUK publisher Charles Knight sought to divert plebeian readers from the radical presses, Murray aimed to compete with such middlebrow publishers as Knight. He did so by reviving and promoting an elitist version of knowledge. In 1829, shortly before the publication of Volume One of *Principles*, Murray began the Family Library. Issued in fifty-three volumes between 1829 and 1834, the Family Library was a series of original, non-fiction works offered to the public at cheapened prices.[58] With the Family Library, Murray became an innovator among publishers. Inexpensive editions of standard works had been available to the public from the late eighteenth century, but the printing of *new* and especially non-fiction works in cheap formats was a development of the 1820s.[59] The Family Library represented Murray's primary response to the challenges and opportunities posed by the new knowledge industry.[60]

Scott Bennett has argued that the Family Library editions were nothing short of 'counter-revolutionary documents', promulgating conservative ideas of political conditions and gentlemanly authority on knowledge across widening class divisions.[61] The first two volumes – biographies of Napoleon and Alexander the Great – addressed the dangers of revolutionary change.[62] The Family Library aimed 'to counteract the Whig-dominated Society for the Diffusion of Useful Knowledge (SDUK), on whose behalf Knight and others were flooding the market for cheap books with inexpensive "libraries" of Entertaining and Useful Knowledge'.[63] In other words, Murray introduced the Family Library into the popular market for knowledge to preempt the domination of that market by Whig reformers, and to assert his own version of gentlemanly, patrician knowledge reform.

While *Principles* was not a cheap Family Library volume, Murray's drive to find in the knowledge movement a niche for conservative, gentlemanly publishing, as manifest in the series, proved pivotal for the publication of *Principles*.[64] Indeed, *Principles* was a logical extension of the series. For one, Murray and Lyell intended the book not for an exclusively scientific readership, but hoped to reach a wider intellectual reading public, much like the public served by the *Quarterly Review*. Second, to extend the readership of *Principles* even further, Murray made sure that the third edition, the first reissuing of the whole work, was published in a cheap format of four small volumes, and priced at six shillings each, similar to the issues of the series.[65] The work would reach not only gentlemanly readers, but also the readers of other cheap works produced in the new knowledge industry.

Lyell was Murray's model author for making scientific knowledge readable for a wider public – those readers who might rarely if ever riffle through institutional *Proceedings* or *Transactions* of the Geological Society, for example.[66] 'There are very few authors, or have ever been, who cd write profound science & make it readable', Lyell later reported Murray as saying.[67] In publishing *Principles*, Murray built on the reputation he himself had helped to establish, extending his own entrepreneurial aims within the new knowledge industry.

Principles and the new science genre

As James Secord has pointed out, the new compendious science treatises that began in this period, and of which *Principles* was an early instance, 'gained prominence because of a market-led demand for synthesis ... It was all very well to focus on strata hunting or stellar mapping within the meeting rooms of the Geological or Astronomical society', but readers of treatises wanted 'general concepts and simple laws'.[68] Lyell's *Principles* would provide such 'general concepts and simple laws' for geology. While reaching a wide readership, Lyell would introduce the general principles according to which observers in the field should pursue their objects.

Lyell had wanted to model the treatise on Mrs. Marcet's popular *Conversations on Chemistry* (1806).[69] Consisting of a series of conversations between a 'Mrs. B' and two young lady pupils, Caroline and Emily, *Conversations on Chemistry* treated various experiments, explaining them in a familiar language suitable to a lay readership. With genteel parlor propriety, Mrs. B. boldly takes on subject after subject relating to chemistry. They talk about Watt's new steam engine, for example. With such conversations, Marcet hoped to enable readers to understand the experiments that she witnessed at the Royal Institution, as conducted by Humphry Davy.[70] *Conversations* became the most popular book on chemistry in the first half of the nineteenth century, and went into sixteen British and two French editions during her lifetime.

Lyell even went so far as to propose *Conversations on Geology* as the first title for his book, but he was preempted in 1828 by an anonymous author.[71] The actual *Conversations on Geology* contained a Mosaic history of the earth, 'just the kind [Lyell] wanted to combat'.[72] As Lyell's book grew into a much more substantial treatise, the title would have been inappropriate anyway. He soon diverged from the 'Conversations' genre. *Principles* became one of the forerunners in what was then new a genre of book publishing – the 'popular' treatise on a particular branch of knowledge, readable by an educated, non-professional public, yet representing a synthesis of the field, and including a new, original theory. It was both popular and groundbreaking science. Such treatises certainly predated *Principles*. But this genre became the dominant model for new works as that other Enlightenment model, the unifying encyclopedic genre, was abandoned.[73]

The establishment of a genre is, of course, predicated upon followers. The new genre burgeoned from the early 1830s. Herschel's *Preliminary Discourse on the Study of Natural Philosophy* (1830), volume one of Dionysius Lardner's *Cabinet Cyclopedia*, was a similar effort at semi-popular science publishing by competing publisher, Longman. *Preliminary Discourse* contained one of the most 'influential evaluations' favorable to Lyell's *Principles*.[74] Longman and the series editor, Lardner, issued 133 volumes in the *Cabinet Encyclopaedia* at six shillings each, including preliminary discourses on natural philosophy, natural history, astronomy, botany, zoology, geology, and other areas of knowledge. Other popular works in specific knowledge areas included the Bridgewater Treatises on 'the Power, Wisdom, and Goodness of God', eight works of natural philosophy commissioned by the Earl of Bridgewater in support of Natural Theology, published by William Pickering; Mary Somerville's *On the Connexion of the Physical Sciences* (1834); Robert Chambers's series, *Introduction to the Sciences* (1838) and *Vestiges of the Natural History of Creation* (1844); and, of course, *On the Origin of Species* (1859).

Murray and natural philosophy

While the political character of the Family Library was conservative, even 'counter-revolutionary', when it came to the publication of works of natural philosophy or by natural philosophers, Murray's publication record was quite ambiguous. From 1830, within the period of a year, he published three works on nature significantly at odds with each other: Humphry Davy's posthumous *Consolations in Travel* (1830),[75] Charles Lyell's *Principles of Geology* (1830–1833), and Thomas Hope's *Essay on the Origin and Prospects of Man* (1831).[76] The three works presented three very different approaches to the topic of progress in nature.

Given the evidence of supposed progress in the history of the earth, including the inorganic, but especially apparent in the fossil series as revealed by geologists, three plausible interpretations were available to naturalists. Theologically oriented naturalists saw evidence of a gradually cooling earth; peopled eventually by serial, special creations 'fitted' by a Designer for their particular environments; and culminating in 'man'. Davy's *Consolations,* essentially a work supporting Natural Theology, proffered this theory, and associated uniformity in geology (as earlier advocated by Hutton and later by Lyell) with species transmutation and evolution. The second possibility was that of a materialist progress narrative, which usually included global evolution and species transmutation. Hope's *Essay,* although written in the language of metaphysics, was a materialist, transmutation narrative. This three-volume tome, filled with philosophical ramblings, refuted Biblical creationism and argued for the evidence of spontaneous generation of organic from inorganic matter, and the development of increasingly complex species from primitive types. Lyell's *Principles* represented the third alternative, which was to argue that any such evidence of progress in the history of the earth, including the fossil record, was simply invalid. His disavowal of progress in nature was intimately connected to his reading of the theory of species transmutation in Jean Baptiste Lamarck's *Zoological Philosophy* of 1809. The book had impressed Lyell, but he reacted very negatively to the prospect of a non-human ancestor for humans, as well as the association of materialism with French revolutionary politics.[77] Arguably, Lyell also reacted to the undercurrent of radical science found in the publications of political and medical radicals in the post-war period. His awareness of the growing audience for such radical science must have urged him to answer the threat embedded in natural progress.[78]

The debate over progress in nature took on political overtones given the connection of progress to materialism and human perfectibility in the works of the French *ideologues* and Encyclopaedists. As Adrian Desmond has argued, the prospect of evolution represented a challenge to the social order, with the implications of an 'uprising nature' signaling a threat of revolution by the lower classes.[79]

It is almost inconceivable that such a savvy publisher as Murray considered works in natural philosophy to be above the sociopolitical fray, especially given the responses to radical science in his *Quarterly Review* in the previous decade. Instead of exclusively considering political implications, however, he may have thought chiefly in terms of the reputations of authors and not the actual character that works might take.[80] In any case, Murray looked for an answer to the programs and publications of the new knowledge industry, and in the process, his record in the publishing of works by naturalists or in topics of naturalism certainly might appear to be one of mixed allegiances, if not for the fact that it took epic form in the publication of Lyell's *Principles*.

The publication history of *Principles* evidences the emergence of secularism within geology as a *reformist* as opposed to a radical, revolutionary intervention in the field. This was accomplished in part by using Lamarckian evolutionary theory as a foil, a radical alternative to be eschewed, in favor of Lyell's essentially cyclic reading of the appearance of species with 'man' as a recent, exceptional case. The objective was to account for the evidence credibly, while preempting radicalism. Lyell accomplished this as he was alerted to the possibility of uniformity's potential association with evolutionary narrative by the posthumously published literary work, *Consolations in Travel* (1830), written by the great chemist Humphry Davy.

Consolations in Travel, Or The Last Days of a Philosopher

Lyell read a copy of *Consolations* in March of 1830, before its publication. In *Consolations*, Davy had associated the notion of uniformity with evolutionary doctrines and Lyell reacted to the manuscript in time to make significant additions to Volume One of *Principles*, which was not submitted in complete form to Murray until the end of June 1830. His emendations to Volume One included his theory of climate change and the refutation of 'the progressive development of organic life', including discounting the evidence for progress in the fossil series.[81]

Lyell was in frequent contact with Murray during this time, and undoubtedly knew about the contents of *Consolations* before it was released to the public.[82] Given his long association with Murray, he likely had seen passages of the book in proof, or even been privy to hand-written manuscript. His non-sequential and misrepresentative quoting of *Consolations* in *Principles* evidences that he was looking at proofs or manuscript, and not the published text.[83] If Lyell had seen an earlier copy that was later revised, the importance of the publishing house connection becomes even greater for content of *Principles*. Otherwise, the 'mistakes' provide evidence that Lyell misquoted for argumentative purposes.[84] For reasons that will become clear, the latter is more likely. In any case, Lyell's reaction to *Consolations*, as discussed below, was very apparent in *Principles*, and certainly depended upon the fact that Murray published both authors' books.

Written while Davy was on his deathbed, *Consolations* is a series of dialogues between fictional characters, intended to be 'ideal' representations of particular belief systems, according to the preface by his brother and biographer, John Davy.[85] The three characters – Ambrosio, Onuphrio, and Philalethes – represent, respectively, 'a Catholic of the most liberal school', a politically liberal English aristocrat (whose 'views in religion went even beyond toleration ... entering the verge of skepticism'),[86] and a cosmic, evolutionary visionary and dreamer, the most impressionable of the three.

The dialogues begin with Philalethes's vision. Philalethes's sensibilities have been heightened by a discussion on the fates of religions and empires while sitting within the Roman Coliseum at dusk. Left alone, he falls into a trance-like state and is confronted by a voice emanating from an intense light. The voice is an 'intellectual guide', a spiritual being he calls 'the Genius', who leads him on a journey through time and space.[87] The Guide is permitted to show Philalethes a glimpse of 'the scheme of the universe',[88] and the laws that regulate it. Beginning with a vision of cave men, he is gradually led on a tour of human history, to the contemporary moment. The Guide conveys a progress narrative of the human race, from savagery to civilization. Human beings begin in a primitive state and continually improve. Upon continual refinement, human souls remove to higher states of being beyond the earth.

In addition to revealing how the improvements in knowledge and technology of one age are retained and passed along to the next, the Genius shows how personal characteristics acquired by peoples are transmitted to their descendants in a method resembling the theory of the inheritance of acquired characteristics as propounded by Lamarck in his *Zoological Philosophy* of 1809:

> In man, moral causes and physical ones modify each other; the transmission of hereditary qualities to off-spring is distinct in the animal world, and in the case of disposition to disease it is sufficiently obvious in the human being. But it is likewise a general principle, that powers or habits acquired by cultivation are transmitted to the next generation and exalted or perpetuated; the history of particular races of men affords distinct proofs of this.[89]

Human beings thus ascend from lower types and evolve into higher types over generations. Unbeknownst to themselves, invaders and conquerors, who promote migration and miscegenation, are 'instruments of a divine plan'. Imperial conquest has a purpose, 'that of improving by mixture the different families of men' – of enlarging the intellects of the 'inferior' races, while strengthening the overly developed and thus diseased sensibilities of the 'superior' ones. Thus, the vision exhibits ethnocentrism and racism, paradoxically common in such 'progressive' theories, as well as the importance of empire and colonialism for progress, soon to be made commonplace with Whewell's Bridgewater Treatise.[90] The overall purpose or 'great object' of such methods of improvement, the Genius explains, 'is evidently to produce organized frames most capable of the happy and intellectual enjoyment of life'. The scheme of the universe consists in continually rarified sensation and thought as beings enjoy an ever-increasing participation in 'infinite mind'.[91]

The Genius adopts a spiritualized middle position between a materialist narrative, and the creationist Biblical one that it contradicts: 'Now, you will say, *is mind generated, is spiritual power created*; or, are those results dependent upon the organization of matter, upon new perfections given to the machinery upon which thought and motion depend? ... neither of these opinions is true ... Spiritual natures are eternal and indivisible, but their modes of being are as infinitely varied as the forms of matter'.[92] On the one hand, the Genius refutes materialism. In holding that mind is generated, that organization is dependent upon matter, and that matter is eternal, materialists are wrong. On the other hand, the Genius also refutes creationism. In believing that God creates both spirit and matter, creationists are wrong. Rather, the Genius declares, spirit is eternal, and it evolves in a manner usually ascribed to matter by materialists.

The tour ends when the Genius is unable to take the visionary beyond the system that the Genius himself inhabits, but assures him that creatures of the Genius's own kind are also continually evolving intellectually and spiritually: 'We are likewise in progression'.[93] The Genius has thus outlined a system of reincarnation or transmutation of spiritual essence in a regular gradation of corporeal-spiritual status.

The subsequent four dialogues recant Philalethes's progress vision. He discusses it with his fellow travelers and is persuaded to adopt a more conventional position, most similar to that of the Biblically conservative Ambrosio. The book concludes by refuting evolution altogether, however spiritualized as it had been in the vision.

The vision and its refutation are important to *Principles*, reminding Lyell of the continued danger associated with evolutionary doctrine, however spiritualized. But the third dialogue, entitled 'The Unknown', directly relates to geology and the question of fossil remains in the strata. It is the dialogue to which Lyell explicitly responds in his *Principles*.

In the third dialogue, Davy represents an amateur naturalist in the person of 'The Stranger'. The Stranger does not 'belong to the modern school of geological schools', the 'plutonic' or Huttonian School, the predecessors to Lyell, who saw 'existing causes' as sufficient for all past geological changes.[94] Instead, his views resemble those of Lyell's boogey men, the contemporary 'Catastrophists'.[95] Most notably, the Stranger associates the Huttonian doctrine of 'existing causes' with the natural descent of species from primitive ancestors:

> You must allow that it is impossible to defend the proposition, that the present order of things, is the ancient and constant order of nature, only modified by existing laws [uniformity], and consequently the view which you have supported [the Huttonian view], must be abandoned. The monuments of extinct generations of animals are as perfect as those of extinct nations; and it would be more reasonable to suppose that the pillars and temples of Palmyra were raised by the wandering Arabs of the desert, than to imagine that the vestiges of peculiar animated forms in the strata beneath the surface *belonged to the early and infant families of the beings that at present inhabit it.*[96]

Thus, the Stranger associates uniformity – 'the proposition, that the present order of things, is the ancient and constant order of nature, only modified by existing laws' – with the view that the 'the vestiges of peculiar animated forms' represent the progenitors, and are the 'infant families' of contemporary species. That is, uniformity implies species transmutation; therefore, it must be rejected. As opposed to this explanation, the Stranger suggests that the fossils of extinct life forms are 'perfect' – that is, complete in themselves and not mere precursors of species to follow. The former skeptic, Onuphrio, finally agrees and recants:

> I am convinced; – I shall push my arguments no further, for I will not support the sophisms of that school, which supposes that *living nature has undergone gradual changes* by the effects of its irritabilities and appetencies [Lamarck]; that the fish has in millions of generations ripened into the quadruped, and the quadruped into the man; and that the system of life by its own inherent powers has fitted itself to the physical changes in the system of the universe. To this absurd, vague, atheistical doctrine, I prefer even the dream of plastic powers, or that other more modern dream, that the secondary strata were created, filled with remains as it were of animal life to confound the speculations of our geological reasoners.[97]

Here, we see uniformity clearly associated with the 'atheistical doctrine' of species transmutation. Of course, the alternatives to it – a 'plastic force', or God burying fossils in the secondary strata to confound humanity – are not suggested as viable alternatives, but as deliberate exaggerations; any explanation for the fossil series is apparently preferable to the explanation by species transmutation. By the end of *Consolations*, all of the skeptical Socratic interlocutors have regulated their views, abandoned skepticism and embraced, more or less, orthodox positions.

According to Wilson, Lyell's reading of Davy's posthumously published book prompted him to develop his theory of climate change, as elaborated in chapters six, seven and eight of Volume One of *Principles*.[98] But the more obvious impact of *Consolations* is seen in Chapter Nine. In Chapter Nine, Lyell suggested that 'a late distinguished writer' had raised one of the 'weightiest objections which have been urged against the assumption of uniformity', that the apparent progress in the fossil series *inveighed against* uniformity.[99] Lyell quoted from dialogue three of *Consolations*, which represented many of the assumptions of the Catastrophist school of geology in 1830. Significantly, the following quotation by Lyell is a non-sequential patchwork of passages drawn from pages 143 to 147 of *Consolations*:

> 'It is impossible', he [Davy] affirms, 'to defend the proposition, that the present order of things is the ancient and constant order of nature, only modified by existing laws – in those strata which are deepest, and which must, consequently, be supposed to be the earliest deposited forms, [specimens,] even of vegetable life, are rare; shells and vegetable remains are found in the next order; the bones of fishes and oviparous reptiles exist in the following class; the remains of birds, with those of the same genera mentioned before, in the next order; those of quadrupeds of extinct species in a still more recent class; and it is only in the loose and slightly-consolidated strata of gravel and sand, and which are usually called diluvian formations, that the remains of animals such as now people the globe are found, with others belonging to extinct species. But, in none of these formations, whether called secondary, tertiary, or diluvial, have the remains of man, or any of his works, been discovered; and whoever dwells upon this subject must be convinced, that the present order of things, and the comparatively recent existence of man as the master of the globe, is as certain as the destruction of a former and a different order, and the extinction of a number of living forms

> which have no types in being. In the oldest secondary strata there are no remains of such animals as now belong to the surface; and in the rocks, which may be regarded as more recently deposited, these remains occur but rarely, and with abundance of extinct species; – there seems, as it were, a gradual approach to the present system of things, and a succession of destructions and creations preparatory to the existence of man'.[100]

The pastiche thus begins with the proposition that cannot be defended – that the present order of things is the ancient order modified by existing laws (the principle of uniformity). The *counter evidence* presented is the existence of fossil remains that show an increasing complexity in organization over time. That is, Lyell suggests that Davy had urged that an apparent progressive order of species development implying species transmutation represented an argument *against* uniformity as such. In fact, as we have seen, Davy had explicitly suggested an *association* of the uniformity of nature with just such a progression. Davy didn't reject uniformity because he accepted a progressive version of species transmutation instead. Rather, he rejected uniformity because for him it necessarily implied species transformism in order to be self-consistent.

Following the quoted passage is a hint of an acknowledgement by Lyell that the passage is actually a conflation and reordering of several passages. 'In the above *passages*', Lyell continued, 'the author deduces two important conclusions from geological data; first, that in the successive groups of strata, there is a progressive development of organic life; – secondly, that man is of comparatively recent origin'.[101] While the latter of these two, Lyell asserted, is true, the former is false. By comparing Lyell's quote to the passages from *Consolations* that I have cited above, we may begin to guess why Lyell may have quoted non-sequentially.

As I have said, the 'weightiest objection' that Davy levied against uniformity was *not* that it failed to account for the variation or change in species over time, as Lyell's patchwork quotation implied. Rather, *Consolations* suggested that to be self-consistent, uniformity had to account for such change *in terms of natural law*. Under a theory of strict and thoroughgoing uniformity that necessarily also embraced the organic world, organic change would occur by means of natural law acting upon organisms to gradually produce new species. As Davy had suggested, uniformity implied 'that the present order of things [including living things], is the ancient and constant order of nature, only modified by existing laws'. If the present order had been merely the past modified by existing laws, then the contemporary species were the descendants of past species modified by existing laws.

Reading Davy's posthumous work, Lyell saw what might become of his 'assumption of uniformity' if the 'progressive development of organic life' were not explicitly refuted.[102] Lyell thus seized on the text 'by the great chemist' and misrepresented it in order to bury the latter's association of progressive or successive development with uniformity. On the other hand, Lyell's main argument against a progressive order in nature consisted of an appeal to exceptions. He argued that only one exception to the rule – that simpler forms precede more complex forms – was 'as fatal to the doctrine of successive development as if there were a thousand'.[103] And such exceptions, he argued, had indeed been found.[104] His own explanation for the appearance of particular species at particular times had to do with climactic conditions. If the same conditions that had accompanied and allowed for particular forms of vegetable and animal life were to recur, so too would that particular vegetable and animal life.

Consolations in Travel thus prompted Lyell to come to terms with and to discount progress in nature in the early editions of *Principles*. (It is well known that Lyell later became an advocate of Darwinian evolution and modified his text accordingly.) In light of Davy's intervention, the renunciation of progress was vital to the introduction of the doctrine of uniformity, that is, if Lyell wanted to avoid the charge of atheistic materialism. Lyell's object was to show that, far from implying transmutation, uniformity opposed progress. The apparent progress in the fossil record was in fact built on feeble evidence and easily overturned. Likewise, to guard against the kind of interpretation Davy had given to the doctrine of uniformity, Lyell excluded both progressive, serial, special creations, and species transmutation, at one and the same time. Far from supporting such beliefs, uniformity, Lyell argued, was their antithesis.

An Essay on the Origins and Prospects of Man

Lyell may have also seen proofs of another work of far more dubious provenance in terms of naturalism: *An Essay on the Origin and Prospects of Man,* written by Thomas Hope, a famous novelist.[105] Murray published Hope's three-volume philosophy in 1831. In tortured metaphysical ramblings, volume one associated a radical empiricism modeled after Hume, with skepticism about the Genesis version of creation. The second volume argued for the spontaneous generation of organic from inorganic nature and developed a progress narrative similar to that of the vision in Davy's *Consolations*.

'Printed in a tiny edition of 250 copies and withdrawn by Hope's executors immediately after publication', its publication was a minor scandal. It was the last evolutionary text published by Murray until Darwin's *Origin of Species* in 1859.[106] Although published after Lyell submitted his first volume, Lyell may have seen proofs of the book, or received reports about its contents in advance. Given its style and authorship, Lyell probably would not have taken the volumes very seriously. Yet the book nevertheless underscored a theme that made troublesome the propagation of any thoroughgoing naturalistic theory, like uniformity: natural progress narratives were 'in the air' in and around 1830, and the prospect of transmutation had to be addressed. The likelihood of Lyell's exposure to the same, and for his taking them seriously, was increased by the fact that even the famous and revered house of Murray was touched by the trend. Thus, the publishing house served as a clearing-house of ideas and arguments that informed and refined his treatise.

Conclusion

This chapter has focused on the context of book production – on the objectives of author and publisher – as they contributed to the knowledge project that culminated with the publication of *Principles of Geology*. First, a consideration of Lyell's earlier articles in the *Quarterly Review* help us to see Lyell's *Principles* as part of a larger project of knowledge reform within gentlemanly circles that began in the late 1820s. In his heralding of the accomplishments and recommending goals for the new scientific institutions, in his reformist ideas for English universities, in his desire to overthrow the Mosaic reading of the earth's surface, and in his attempt to ward off the threat of transmutation theory, Lyell hoped to delineate a new, more professionalized and specialized knowledge class. The principles of geology that Lyell advanced happened to accord well with the kind of knowledge producer that he envisioned; such a producer was neither supported by the Church, nor did he propound ungentlemanly, materialist cosmogonies. He was able to lecture and write professionally, earn a living and gain prestige, and, unlike the literary man, would not be unduly subject to the marketplace. The lack of institutional support was certainly a factor that prompted Lyell's book, both as part of a project to create such a class, and as a means of financial support in lieu of it.

Second, we have seen how a publisher's motive for furthering his own agenda in the new knowledge industry was instrumental in the production of the book. Economic opportunity, a spirit of competition, and a desire to establish gentlemanly control over a vast knowledge industry compelled Murray to enter into a market already opened up by middle- and low-brow ventures. Other reformers and publishers had taken the lead and set the agenda. Simple reactionary refusal to engage in the industry may have suited a few gentleman authors, but it would not serve a publisher with market concerns. Murray had groomed the young gentleman science writer, Charles Lyell, in his *Quarterly*. His science writer had already advocated reforms with the careful, judicious and rhetorical mastery of an experienced, well-heeled barrister. By means of such a careful and gentlemanly writer as Lyell, Murray sought to gain a foothold in the new knowledge industry, without overtly offending his established clientele.

Next, we have seen how a publisher served as a clearing-house of ideas for Lyell's science. Given his intimate connection to the house of Murray, Lyell was alerted to texts – whether in manuscript or published form – which he might have not otherwise seen in advance of publishing his first volume. Alarmed by Davy's treatment of uniformity in *Consolations*, Lyell responded quickly by emending his own arguments and adding a polemic against progress in nature. To avoid being associated with such materialist progress narratives as Davy condemned (and as appeared in Thomas Hope's *An Essay on the Origin and Prospects of Man*), he added last minute addenda to his notion of uniformity. Thus, we see that the publishing house had an enormous impact on the making of Lyell's science.

Finally, and most importantly for the broader argument here, the project culminating in Lyell's *Principles* represents an early division in knowledge production according to secular aims. Although Lyell strenuously disavowed the Lamarckian evolutionary theory that would be embraced by later secularists, he nevertheless etched a line of demarcation that would serve evolutionary theory in the future, and would promote a secular outlook necessary for the establishment of the scientific naturalism later in the century. We also see that, like the scientific naturalism of the second half of the century, Lyell's secularism was tied to professional and cultural ambitions and used as a means for the expression of cultural authority within the domain of science.[107] Furthermore, Lyell's uniformity was both reformist and preemptive. His uniformity was proffered as a trope representing gradualist reformism as a necessary alternative to a radical materialism and atheism of such radical science and politics advocates as Richard Carlile. Lyell's uniformity thus should be seen as a companion to the gradualism of such creeds as Comte's Positivism and as a precursor to Holyoake's Secularism, the latter of which would become vogue amongst British artisan and literary radicals by mid-century.

3

Holyoake and Secularism: The Emergence of 'Positive' Freethought

The facts about George Jacob Holyoake's founding and early leadership of Secularism have been well documented. Yet, to date, Secularism has not been appreciated for its significance in terms of general secularism or modern secularization. In this chapter, I aim to explore Secularism not so much for its success or lack thereof at converting religious believers to its fold, or in terms of its institutional structures and organizational apparatuses. Social historians of Secularism, especially Edward Royle, have done well to demonstrate in great detail such social facts about Secularism.[1] Instead, my focus will be on Secularism as a particular cultural and intellectual formation or constellation. I am interested in Secularism as a historic signpost for what it can tell us about the configuration of the elements involved in its construction, the secular and the religious, and how these interacted with each other and other factors under Secularism's banner. My approach to the structure of Secularism thus lies somewhere between cultural history, intellectual history, and cultural critique. I look for what Secularism can reveal about its historicity as a cultural, intellectual, religious, and social development and what that development suggests about the configuration of what we now understand as modern secularity.

Most studies of Secularism have either ignored Secularism's relation to secularization and the development of a more general modern secularism or secularity, or regarded it as unimportant.[2] Remarkably, in the preface to his second major study on freethought, *Radicals, Secularists, and Republicans* (1980), Edward Royle announced decisively that 'one thesis of this book is that it [Secularism] had little in practice to do with modern notions of the secular'.[3] He also noted that he had not paid much attention to the question of freethought in the higher classes. That is, Royle treated Secularism apart from the better-known 'crisis of faith' phenomenon. But more importantly, he argued that the first movement ever to adopt the name 'Secularism' was irrelevant to what was, in the early 1980s, taken for granted – the notion of (a progressive) secularization as a condition of modernity. Royle illuminated much about this formerly neglected terrain. Yet, while denying Secularism importance in terms of modern notions of secular (or more likely, side-stepping the issue altogether), Royle's histories may have contributed to Secularism's subsequent isolation over the past thirty plus years. Drawing on Royle's *Victorian Infidels* (1974) for 'The Attitudes of the Worker', a chapter in *The Secularization of the European Mind* (1975), Owen Chadwick analyzed the place of Holyoake's Secularism within the broader trajectory of secularization.[4] Chadwick's attitude toward working-class intellection was condescending at best, however. More to the point, he assessed Secularism as an isolated phenomenon that had a limited effect and no real significance for the secularization of the broader culture, a process that he insisted on throughout the book. A similar discounting of Secularism also appears in recent general accounts of secularism. In the monumental *A Secular Age*, which includes a capacious treatment of the nineteenth century, Charles Taylor does not even mention the movement of nineteenth-century Secularism, the first ever use of the term, or its founder, George Holyoake.[5]

Secularism has not always been neglected or otherwise rendered utterly insignificant in broad accounts of secularism or secularization, however. For Graeme Smith, the story of Secularism is one of 'failure', but it is a failure with historical and conceptual importance. In *A Short History of Secularism* (2008), Smith asserts that the 'failure' of organized secularism demonstrates 'that the designation of the "secular" when applied to Western society is not meant to describe people's atheist commitments'.[6] The relative impotence of organized secularism in overcoming religion, he continues, indicates that the term 'secularism' should be redefined. By this, Smith seems to suggest that by 'secularism' or the condition of modern secularity, we might instead be referring to something like the continued co-existence of the secular and the religious. While Smith diminishes Holyoake's contribution and the movement of Secularism as a whole, and furthermore misses the fact that Secularism proper was *not* originally established to 'overcome religion', his argument for a new conception of the 'secular' and 'secularism' accords well with my argument here and throughout this book.

It appears necessary to note that with the term 'secular', Holyoake did not signify the absence or negation of religion, but rather indicated a substantive category in its own right. Holyoake imagined and fostered the co-existence of secular and religious elements subsisting under a common umbrella known as Secularism. For Holyoake, the secular and religious were figured as complementary and co-constituting aspects of what we might now call an overarching secularity but which Holyoake called Secularism. Of course, this understanding of Secularism is at odds with the standard secularization thesis according to which religion is progressively eliminated from the public (and eventually the private) sphere. I argue that Holyoake's Secularism should be understood as an early intervention into what has since become known as modern secularization theory and consider it an historic moment, as the previously unrecognized inauguration of modern secularity.

I begin with a brief introduction to the conception of Secularism, and continue by tracking Holyoake's periodical and pamphleteering career in the 1840s. I distinguish Holyoake from another prominent freethinker, Charles Southwell, and show how Holyoake eventually developed Secularism as a moral program – to escape the stigma of infidelity, but more importantly to move freethought toward a positive declaration of materialist principles as opposed to the mere negation of theology. I then show how Holyoake's Secularism eventually arose from the class conciliation between artisan-based freethinkers and middle-class skeptics, literary radicals, and liberal theists, and as a branch of Secularism distinct from that led by Charles Bradlaugh. I conclude with remarks on the implications of Holyoake's Secularism in connection with modern secularism and our conceptions of modern secularity.

The inception of secularism, in brief

In the late 1840s, a new philosophical, social, and political movement evolved from the radical tradition of Thomas Paine, Richard Carlile, Robert Owen, and the radical periodical press. An innovation of the artisan freethought[7] tradition of which Carlile had been the leading exponent in the 1820s, this movement drew from the social base of artisan intellectuals who came of age in the era of self-improvement, the diffusion of knowledge, and agitation for social, political, and economic reform. The movement was called 'Secularism'.[8] Its founder was George Jacob Holyoake (1817–1906).[9] Holyoake was a former apprentice whitesmith turned Owenite social missionary, 'moral force' Chartist, and leading radical editor and publisher. Given his early exposure to Owenism and Chartism, Holyoake had become a freethinker. With his involvement in freethought publishing, he became a moral convert to atheism. But his experiences with virulent proponents of atheism or infidelity and the hostile reactions to infidelity on the part of the state, church, and press induced him to develop in 1851–1852 the new creed and movement he called Secularism.

In retrospect, Holyoake claimed that the words 'Secular', 'Secularist', and 'Secularism' were used for the first time in his periodical the *Reasoner* (founded in 1846), from 1851 through 1852, 'as a general test of principles of conduct apart from spiritual considerations', to describe 'a new way of thinking', and to define 'a movement' based on that thinking, respectively.[10] In using these new derivatives, he redefined in positive terms what had been an epithet for the meaner concerns of worldly life or the designation of a lesser state of religiosity within the western Christian imaginary. His bold claims for the original mobilization of the terms are corroborated by the OED.[11]

It is important to distinguish Holyoake's brand of Secularism from that of his eventual rival for the leadership of the Secularist movement, Charles Bradlaugh. Unlike Bradlaugh, for Holyoake the goal of freethought under Secularism was no longer first and foremost the elimination of religious ideology from the public sphere.[12] Holyoake imagined Secularism as eventually superseding and superintending both theism *and* atheism – from the standpoint of a new scientific, educative, and moral system. Holyoake insisted that a new, secular moral and epistemological system could be constructed alongside, or above, the old religious one. At the same time, Holyoake welcomed religious believers into the Secular fold. On the other hand, against Holyoake's assertions, Bradlaugh maintained that the primary task of Secularism was to destroy theism; otherwise the latter would impede the progress of the new secular order.[13]

Mid-century Secularism thus represents an important stage of nineteenth-century freethought – an intervention between the earlier infidelity of Carlile and 'Bradlaugh's rather crude anti-clericism and love of Bible-bashing'.[14] While this new movement inherited much from the earlier infidelity of Carlile and Owen, Holyoake offered an epistemology and morality independent of Christianity, yet supposedly no longer at war with it. Had Holyoake's Secularism amounted to nothing more than this, it would nevertheless represent a significant historical development. Yet, mid-century Secularism is also significant in terms of the development of modern secularity, as it is now understood.

Freethought from 'infidelity' to moral philosophy

Given that the rhetoric, arguments and legal battles of the 1840s infidels resembled those of Richard Carlile and his immediate followers of the 1810s and 20s, the later freethought movement has often been seen as an extension of earlier freethought.[15] But the conflation of the two periods glosses over the historical peculiarities of 1840s unbelief, as well as the distinctiveness of the freethought publications themselves. As J. M. Robertson wrote several decades ago:

> It is inevitable that in a progression by way of recording debate and polemic, criticism and resistance, the advance of thought should often be figured as a continuous battle, in which one flag steadily gains against another. But the resultant conception, for the reflective student, is one of perpetual transmutation, the flags themselves, so to speak, being progressively re-made, as the issues are reconsidered.[16]

Setting aside Robertson's narrative of necessary progress, we can see that the fabric of freethought was rewoven in the 1840s. First, the terms of publishing had changed. The freethought movement of the 1840s had benefited from the struggles of the earlier radical pennies, and also, while in some sense a radical rejection of it, the diffusion of knowledge movement that began in the 1820s. The partial victory in the 'War of the Unstamped' had by 1836 already widened the horizon of the cheap periodical. For advocates of a completely 'free press', the ultimate horizon of victory was yet far off. But for publishers and journalists, a new stage in the expression of opinion was dawning. The union of reformist and radical publishers and editors for the removal of the stamp had fostered an environment of cooperation, discussion and debate, as opposed to '*ad hominem vitriol*'.[17] Free inquiry and expression, independent of what might be investigated or expressed, had become the primary object of many propagandists. A respondent to Carlile expressed the new sensibility in 1839:

> Sir – although our opinions respecting the best plan of bringing about the reform we both desire differ very widely, it is at least evident that we are agreed on one point, viz. the paramount necessity of 'Free Inquiry'. So long as this is the watchword, I care not whether any opponents be Orthodox or Infidel, Deist, Demonist or Atheist, Whig, Radical or Tory, because there is an obvious *desire* for truth.[18]

Free inquiry itself became the general object of the freethought movement as well, but its *particular* object was the free expression of atheism or materialism. Thus, vituperation continued to characterize the rhetoric of some of its spokesmen for the early part of the 1840s.

Second, the material conditions of the working classes had worsened, and, with the failures of first wave of Chartist political reform and

Owenite socialist schemes, 'the spread of atheism increased sharply after 1840, as disillusioned socialists adopted a more militant stance'.[19] At least, that is, a rhetoric of atheism *initially* became more pronounced. The conciliatory political and religious stances adopted by Robert Owen and *The New Moral World* drew harsh criticism from the infidelity segment of the socialist movement.[20] While Owen and the Socialists of the *New World Moral World* had hoped to effect a rapprochement with religious and state authorities, those who had been attracted to the infidel more than to the socialist aspect of Owenism drew on earlier sources, cobbling together the remnants of the tradition of Paine, Carlile, and the infidelity of Owen, to promote a newly reinvigorated radical infidel freethought movement.[21] In 1841, the former Owenite Social Missionary, Charles Southwell – with Maltus Questell Ryall, 'an accomplished iconoclast, fiery, original, and, what rarely accompanies those qualities, gentlemanly', and William Chilton, a radical publisher and 'absolute atheist' – founded in Bristol a periodical that its editors claimed was 'the only exclusively ATHEISTICAL print that has appeared in any age or country', *The Oracle of Reason, or Philosophy Vindicated.*[22]

Charles Southwell might, with important exceptions, be thought of as the Ludwig Feuerbach of British infidelity in the early 1840s, at least as Marx characterizes the latter in *The German Ideology* (1845).[23] In this work, contemporaneous with the founding of *The Reasoner* (1846), Marx argued that the Young Hegelian Feuerbach was merely substituting one kind of *consciousness* for another, 'to produce a correct consciousness about an existing fact; whereas for the real communist it is a question of overthrowing the existing state of things'.[24] Marx argued that as warriors against religious concepts for the purposes of human liberation,

> [t]he Young Hegelians consider conceptions, thoughts, ideas, in fact all the products of consciousness, to which they attribute an independent existence, as the real chains of men [...] it is evident that the Young Hegelians have to fight only against these illusions of consciousness. Since, according to their fantasy, the relationships of men, all their doings, their chains and their limitations are products of their consciousness, the Young Hegelians logically put to men the moral postulate of exchanging their present consciousness for human, critical or egoistic consciousness, and thus of removing their limitations.[25]

An atheist martyr, the criticism cannot be applied to Charles Southwell without qualifications. His writing constituted a political act with material and political consequences. While writing of 'practical rights' with 'practical powers', as opposed to 'abstract rights', which were 'mere chimeras',

Southwell wanted to prove his rights in actual practice. However, the end he hoped to effect was in fact a revolution in ideas, which would, he thought, eventuate a change in material circumstances – precisely what Marx critiqued in Feuerbach.[26]

My aim is not to engage in an extended comparison of English infidelity and post-Hegelian German philosophy, but rather to underscore the irony of Southwell's abstraction of atheistic materialism from its socio-historical context in order to contrast it with the direction freethought was soon to take under Holyoake. Despite his polemical engagements with Owenite socialists on matters of atheist principles, his reports on his legal case, and his querulous letters from jail, one might argue that Southwell nevertheless warred on the level that Marx referred to as ideological, seeing religious ideas as the 'chains of men'. Southwell gave the sense of atheism as a purely intellectual affair, as the proclamation of a truth that has arisen at different times in places, including ancient Greece, but that has been continually thwarted by priests of all ages:

> Space is something or nothing, a reality or a fiction, that which really exists, or a negation of all existence; if the former it cannot be a god that Christians will accept, for that which is real must be corporeal: but they reject a matter god and will not agree with the Stoics, that god is a divine animal: if the latter, that is, if those who will have it that space is god, are driven to admit, as they necessarily must, that space is the negation or absence of matter, an absolute nothing, why, then, we fall upon the *ex nihilo nihil fit*: Englishished – out of nothing nothing can come. As plain a truth as any to be found in Euclid. Which makes the question stand thus: if space is an actually existing something, it must be matter; but that a matter god is no god at all, is allowed by the Christian world. In the second place, space cannot be a god, if it signifies pure emptiness or absence of matter, because the absence of matter, could it be conceived, is a nothing; and to refine god into nothing is to destroy the idea of such an existence, and to proclaim that Atheism we are *labouring to teach*.[27]

Soon growing impatient with the lack of response to his philosophical disquisitions,[28] however, Southwell opened the fourth number of the *Oracle* with a caustic and belligerent article entitled 'The Jew Book'. Here, he took aim at the sacred text, which proved more dangerous and thus more effective for his purposes.

> That revolting odious Jew production, called BIBLE, has been for ages the idol of all sorts of blockheads, the glory of knaves, and

> the disgust of wise men. It is a history of lust, sodomies, wholesale slaughtering, and horrible depravity, that the vilest parts of all other histories, collected into one monstrous book, could scarcely parallel! Priests tell us that this concentration of abominations was written by a god; all the world believe priests, or they would rather have thought it the outpouring of some devil! [29]

As James Secord notes, Southwell's polemic may be regarded as an 'ugly attempt to exploit popular anti-Semitism to mock the Bible'.[30] Southwell later admitted in his autobiography *Confessions of a Freethinker* that he had purposively written to provoke the authorities.[31] On the date of its publication, Southwell was arrested for blasphemy and taken to Bristol jail.[32] His trial became a *cause celebre* in the liberal press.[33] His self-defense was unsuccessful, however, and on 15 January 1842, he was fined 100 pounds and sentenced to a year's imprisonment.[34]

With Southwell incarcerated and unable to manage the publication, George Jacob Holyoake became the editor of the *Oracle*. Under Holyoake's editorship, a change in rhetoric was immediately evident. Holyoake would not change the *Oracle*'s purpose – to 'deal out Atheism as freely as ever Christianity was dealt out to the people'[35] – but he refrained from such odiously provocative and offensive denunciations as Southwell's 'The Jew Book', moving the mission of the *Oracle* toward a positive declaration of atheistic and materialist principles, and away from a mere negation of theism.[36] Cleric baiting and Bible roasting were eventually replaced by more eloquently impassioned pleas, exemplifying a principle of free speech without an ethic of vitriolic attack. Eschewing incendiary rhetoric, Holyoake sought sympathy for atheism on the basis of the conditions of poor workers and the failure of the Christian state to remedy them. Like Thomas Cooper, the Chartist poet and leader in Leicester, Holyoake saw Christianity as irrelevant to the suffering of the poor, and although not as depressive, like John Barton in Elizabeth Gaskell's Hungry Forties novel, *Mary Barton* (1848), he was 'sadly put to make great riches and great poverty square with Christ's Gospel'.[37] His loss of faith had been occasioned by moral repugnance over the apparent indifference of the Christian state to the conditions of the suffering majority. Holyoake's was first and foremost a moral conversion to atheism.[38] As he stated in a lecture in Sheffield on 'The Spirit of Bonner in the Disciples of Jesus', 'the persecution of my friend [Southwell] ... has been, within these few weeks, the cradle of my doubts and the grave of my religion. My cherished confidence is gone, and my FAITH IS NO MORE'.[39]

In the third number as interim editor of the Oracle, in 'To What Do Things Seem Tending?' – written in the 'signs of the times' periodical genre inaugurated by Thomas Carlyle – Holyoake elaborated the intimate relationship between a materialist, atheist philosophy and the material conditions that might engender such a philosophy:

> Want has made more converts than preaching, of late days. Gospel and good dinners did very well together, as fat old abbots, and rubicund-nosed parsons, can tell. Christ and a crust, merely, never in this world went down well, in spite of all that pious tracts say to the contrary. But Christ, without the crust, people soon die upon, as poor-law guardians and relieving officers can, and do abundantly testify.[40]

During his childhood, Holyoake's sister had died while his mother was away from home paying the Church rates and Easter dues. Conditioned by this personal loss from material want and its connection to religious observation, Holyoake had been predisposed to lose his faith in divine providence. While serving a sentence for blasphemy in Cheltenham Jail from 1841 to 1842, Holyoake's daughter died, likely for lack of medical attention owing to lack of funds. His continual exposure to worldly want and suffering eventually spelled the end of whatever faith he may have had.[41]

As Holyoake saw it, want and knowledge were collaborators vying against superstition for control of the mid-century mind. 'With the progress of knowledge, spirit and spiritual things have evaporated like ether poured out in the sunbeams'.[42] Spirituality was a mirage that might have been utterly eradicated by knowledge, but because knowledge was, like Prometheus, still 'changed to the rocks of superstition, and plucked at by the vultures of theology ... suffering teaches lessons where reason could not impart truth'.[43] What reason could not do in the bidding against religious superstition, the social conditions were accomplishing by deprivation:

> 'Of thirteen children only one is left (said a poor old woman to Alderman Kelly, the other day in Guildhall), and she is transported; I have travelled here from Hunslet, to see her for the last time; see my nakedness and rags (stretching out her gaunt withered and bony arms before the court): father, mother, brother, sister, children, all gone; I have no friend left but god, and I begin to think he is rather hard upon me in my old age'. Misery had done its work; groundless piety was expiring, where it evidently had been most tenaciously cherished.[44]

Holyoake acknowledged that the diffusion of knowledge and the spread of powerful ideas were not always sufficient to win converts to materialism. Consistent with his Owenite roots, he considered belief a product of circumstances. And as in Carlile's rationalistic empiricism, knowledge meant 'familiarity with the knowable, the avenues of which are the senses'.[45] Knowledge could come from study, or it might be produced by extraordinarily stringent sensations; in fact, it might require them. If the diffusion of knowledge had failed to drive out superstition, the misery of poverty would not. However, without the spread of knowledge, unbelief might arise only after it was too late to do any good. The diffusion of knowledge was likewise necessary to promote unbelief in order that material conditions might be improved.

While the *Oracle* still retained remnants of the old infidelity, Holyoake and company had squarely shifted the focus to what is known as 'the Condition of England question'.[46] By the early 1840s, the freethought radicals had integrated what has been termed 'the new analysis' into their loosely assembled 'program' of reform. The 'old analysis', an extended attack on 'Kingcraft and Priestcraft', on taxes and sinecures, which also encompassed Republicanism, or alternatively the rhetoric of universal suffrage, gave way under the new context of advancing industrialism. While retaining something of the old analysis, the new analysis drew on the work of Thomas Hodgskin, Charles Hall, and Robert Owen to include a criticism of the competitive system of economic exploitation and of political economists, especially Thomas Malthus, as its primary apologists.[47] The Hungry Forties had done for materialism what a war of ideas never could, and as if validating Owenite doctrine, the force of circumstances made for the birth of a new emphasis. Loosely affiliated with Owenism and Chartism, freethinkers sought to develop their materialist convictions into 'Political and Social Science',[48] a class-based critique of Malthusian political economy and an alternative political and social science. By 'science', they meant knowledge and action producing observable and predictable changes for the improvement of material conditions. Given its movement toward political and social science, freethought was set to enter a new 'constructive' stage.

The Movement: the 'third stage' of freethought

When Southwell declined to resume editorship of the *Oracle* upon his release from Bristol jail, Holyoake and company decided to fold the publication; yet they were committed to keeping freethought publishing alive. *The Movement And Anti-Persecution Gazette* was founded on

16 December 1843, allegedly to continue the mission of the *Oracle* and to report the activities of the Anti-Persecution Union.[49] Assisted by Ryall, Holyoake would be the primary editor and contributor. Central to the *Movement* was its departure for freethinking journalism. Not only did the editors maintain the tonal and rhetorical moderation characteristic of the *Oracle* after the removal of Southwell but also the *Movement* launched the 'third stage' of freethought. As Holyoake saw it, the first two stages, free inquiry and open criticism of theology, were essential, but not constructive. They sought to free expression and to destroy religious ideology. The third stage, however, involved the development of morality: 'to ascertain what rules human reason may supply for the independent conduct of life ...'.[50] The difference in emphasis marked what Holyoake later referred to as the 'positive' side of freethought, which would not simply destroy theism, but replace its moral system with another. With this, Holyoake echoed Auguste Comte, who held that 'nothing is destroyed until it has been replaced'.[51]

The *Movement* was perhaps the first freethought periodical in Britain to emphasize a predominantly constructive approach, considering its duty to be to work toward the improvement of the conditions of the working classes, adopting to the circumstances of the Hungry Forties the Benthamite motto – carried as the epigraph following the title of every number – 'to maximize morals, minimize religion'.[52] Inaugurating the development of a liberalized epistemological and moral system independent of theology and relying on a rational application of methods derived from the observation of society, the *Movement* began an undertaking parallel to the positivism of Auguste Comte in France, while anticipating the social and political philosophy of John Stuart Mill's *On Liberty* (1859).

The weekly introduced a materialist epistemology as the basis for thought and action: 'Materialism will be advanced as the only sound basis of rational thought and practice'.[53] As has been noted by historians, 'materialism' was a polysemous designation in the period. By materialism, Holyoake meant a reliance on material means for material change and the improvement of real conditions in 'this life'. While Holyoake would continue for a time to reiterate the standard arguments for atheism and materialism, he would do so under the new head of 'naturalism'. In a serial article entitled 'The Principles of Naturalism', the doctrine of materialism was elaborated. Materialism, naturalism and atheism were essentially equated, differentiated only by context. 'Naturalism signifies a system of reasoning and belief founded on known phenomena. Philosophers style it Materialism, and religious people call it Atheism'.[54] In equating the three terms, Holyoake could begin to alternate between them, depending on the rhetorical requirements of the situation. His main objective, however, was to introduce the 'naturalism' as a replacement for atheism, a step in a strategy that would prove determinative for the eventual development of Secularism. The term 'naturalism' was somewhat disarming. Atheism denied God while naturalism merely suggested an exclusive attention to natural phenomena. The strategy allowed freethought to position itself in terms that could more easily be applied to scientific and moral reasoning, without invoking metaphysical questions.

For many freethinkers, materialism had also included the denial of anything other than the material world; but denial of the extra-material became inessential for Holyoake. He strongly declaimed the neglect of material conditions on the basis of 'spiritual' considerations, yet as his thinking developed, he would come to refuse to speculate on the existence of a non-material reality. While not denying the extra-material, Holyoake deemed it irrelevant to 'science' and progress. That is, by the late 1840s, Holyoake's was already becoming a *methodological* as opposed to an ontological or metaphysical materialism. His development of a methodological materialism may be seen as a precursor to Huxley's agnosticism, and the scientific naturalism to which the latter was central. (See Chapter 4.)

Yet unlike Southwell before him, or Huxley after him, Holyoake wanted to avoid philosophical complexity, suggesting that success of a creed depended on the simplification and popularization of principles:

> All men do not view the question of the existence of deity with the profundity of Spinoza, or the meditative acumen of Strauss. The majority of mankind regard chiefly its bold and broad features. This may be less erudite, but it is more practical. No question is properly examined until its bolder traits are described; nor fully understood until mankind have been able to view them. It will, therefore, be useful to the present question in its most obvious bearings, and for this purpose the presumptive evidence against the belief in deity must first be offered to the reader's notice.[55]

Arguing the 'presumptive evidence' against the existence of deity, the primary objections to theism involved not its logical fallacies or inconsistencies, but rather the social and political effects of belief. He listed four such presumptive evidences, the last of which was 'its moral influence; for the most atrocious monsters have always been its most zealous supporters, and its professors the most enduring foes of liberty, virtue, improvement and truth'.[56] The disproof of belief was in the pudding.

In an article entitled 'The Mountain Sermon', the freethought version of the Sermon on the Mount, Holyoake sketched what was later to become part of Secularism's moral system, based upon the reasoning of humanistic Benthamite Utilitarianism:

> The business of mankind is with man, and it is unwise to make human action, relative to, not earth, but to heaven, and imaginary consequences hereafter. The best source of the affections is to be found in themselves, and the result of reason, which sees its own happiness in the happiness of the greatest number. The alms-giving of the sermon on the mount, has no one worldly object in view, but promises a greater reward, in the estimation of believers, than any can receive on earth. True charity only looks to the justice of the donation, and cares little whether it be known or not, except as the publicity of the good act may induce others to do the same. Generosity contemplates no other reward, than the good which was the object of the donation; but, if I part with my money from a preconceived idea that I get a reward in heaven, or here on earth as a return from heaven, that was the object, and not the good of mankind ... When the object of charity is heaven, and not man, almsgiving will look more to the salvation of the soul, than the benefit of man. Money will go only to the faithful, or the conversion of infidels. Man, as man, is entirely overlooked in these sentiments of Jesus, as human nature is everywhere, in his mouth, made a sacrifice to heaven.[57]

We see that freethought had clearly moved beyond the mere denunciation of 'Priestcraft' and a denial of deity, and had begun to describe a humanist and materialist basis for benevolent human action. As such, the above passage anticipates Secularism, which was introduced in *The Reasoner* and invented, in part, to replace or superintend the moral system of Christianity with one within which actions were considered in terms of the benefits accruing to 'this world' alone. It also would allow for the pursuit of positive knowledge without reference to natural theology.

The Reasoner: the upward mobility of freethought

The Reasoner was founded by Holyoake with the fifty pounds he won for his five entries into the Manchester Unity of Oddfellows contest for the best new lectures, to be read to graduates into the Oddfellowship.[58] The publication became the central propagandist instrument for freethought. By the time he began the new weekly, Holyoake was now a leading freethinker. His earlier position as an Owenite Social Missionary, his well-publicized trial for blasphemy, his secretariat of the Anti-Persecution Union, and his editorship of the leading freethought journals, had secured his reputation.[59]

The Reasoner became the longest-standing freethought publication of its time, publishing from 1846 to 1861. An eight-page weekly like its forebears, the first number appeared on 3 June 1846, at a price of two pennies, a price exceeding that of the earlier two periodicals, which had ranged from a penny to a penny and a half. The first series ended in 1861, soon after the introduction of Charles Bradlaugh's *National Reformer* in 1860. The cessation of the first series marked the effective end of Holyoake's singular prominence in the Secular movement, as Bradlaugh, the 'Iconoclast', founded the National Secular Society in 1866, and became its first and long-standing President, until 1890, a year before his death.[60]

Of the publications with which Holyoake had been involved to date, he often suggested that *The Reasoner* most nearly characterized his own ideas, style and rhetoric, although these changed over time.[61] As he wrote in 1847:

> A great number of people look at me through the eyes of the *Oracle of Reason*, although I never had the personal control of that paper a single week, being always in a distant part of the country from the place of publication. It bore my name when you [Paul Rodgers] knew it. I was legally responsible for it. That was the part devolving upon me. Yet, when the originator of the paper abandoned it, I, with other friends, stood by it, although I had to protest against some of its contents.

Holyoake wished to distance himself from the *Oracle* and Southwell in particular, claiming that the periodical had never been under his control, despite his nominal editorship. Southwell responded, objecting to Holyoake's characterization of his own 'abandonment' of the paper and criticizing what he termed Holyoake's 'run-with-the-hare-and-hold-with-the-hounds policy' and 'writing down the *Oracle*' in order to earn 'empty compliments from learnedly ignorant Christians'.[62] But Holyoake's protests were chiefly against the rhetorical excesses in denunciation practiced in the *Oracle* and demonstrated by Southwell's letter. William Chilton, who remained a contributor to the *Reasoner*, also voiced his 'regret' for the 'coarseness, vulgarity, and even brutality' in his own *Oracle* compositions, although he felt the language justified by the circumstances of the imprisonment of most of his atheist friends.[63]

Holyoake was not only interested in distancing himself from Southwell's rhetoric, but he also had another kind of freethought movement in mind. After avowing atheism, he was never to alter his views or disguise his beliefs, but his tactics for affecting his goals were quite flexible, to say the least. Often acerbic in his assertions of atheism, he nevertheless chided atheists who alienated liberal theists. While maintaining his right to the profession of atheism, he came to advocate the accommodation of other than atheistic views within a broader movement. Unbelievers, deists, monists, Utilitarians, and liberal theists might all cooperate, provided that together they promoted a morality, politics, economics, and science of worldly improvement. While a seemingly contradictory position that alienated and angered some, it represented the differentiation of a religious public sphere, within which belief and unbelief coexisted by means of an overarching secularity. Secularism thus marked a new stage in secularity itself, evincing a recognition that religious belief was unlikely to disappear.

Holyoake's accommodation and admission of theism and theists was to become a chief source of his conflicts with militant freethinkers, both of his own generation, including Southwell, and the next, as represented by Charles Bradlaugh. Bradlaugh would later argue that Secularism meant an uncompromising advocacy of atheism. He vehemently denounced the accommodation of belief and believers in Holyoake's *Reasoner* and Secularist Society and later, with his militant bravado and readily digested declamations of the Bible, won leadership of the larger wing of the movement.[64]

Yet Holyoake's objectives and approach should be understood in terms of the gains made by earlier freethought, as well as the challenges and opportunities posed by mid-century circumstances. By the 1850s, freethought had already won a wider toleration, given the legal battles of the previous decades. A growing distaste for prosecuting atheists came with the partial defeat of social prejudice against heterodoxy. Freethought had already made incursions into 'respectability' by the late 1840s. As Holyoake saw it, freethought now faced new challenges if it was to spread beyond its narrow margins. Its associated publications having already made its case in the most 'extravagant' terms, the *Reasoner* could now promise refinement in keeping with middle-class tastes.[65] The new tone and approach were only justified by the increased toleration of the times, owing largely to freethinkers' earlier encroachments on propriety:

> having once *actually* trodden on the tender excrescences of error we promise the very best *amende* in our power ... The greatness of our cause is no longer obscured by the extravagance of party zeal. We are not now goaded, nor fretted, nor chafed by contumelies, but born in the transmitted spirit of freedom and social reformation, we have reflected on the past and calculated on the future, and coolly estimating the worth of the objects we seek, we are willing to hazard much to gain them. This deliberate resolve is not to be confounded, in its prospective bearings, with the angry impatience of overburdened men.[66]

That is, by the time that Holyoake and company founded Secularism in 1851, a space had already been cleared for it in the public sphere. Moreover, within this discursive, social space, Holyoake soon found collaborators drawn from the middling ranks.

In volume five of the *Reasoner*, Holyoake began a serial article entitled 'Rudiments of Public Speaking and Debate, or Application of Logic'.[67] In this disquisition on rhetoric we can make out an important facet of Holyoake's political objectives. He began to establish some distance between himself and his artisan roots, with frequent references to 'the people' and 'the multitude', whose lack of knowledge he apparently did not share. He suggested a graduated 'progression' toward 'the highest results of philosophy', beginning with those who have only just become 'sensible of their ignorance', and who 'are now engaged in a double battle against Want and Error'. But knowledge is not attained by all of humanity all at once. Rather, humanity is uplifted in 'a series of stages', with 'individuals first, then groups, then classes, then nations are raised'. The mode of address placed Holyoake amongst the first individuals in his group and even of his class to be elevated so as to address the next group in line, 'the class of young thinkers to whom knowledge has given some intellectual aspiration, and fate denied the means of its scholastic gratification'. Like himself, having been students at 'Mechanics' and Literary Institutions', which 'cannot *cultivate* their frequenters' but only '*stimulate* their improvement', this class was neither the 'elementary nor the ultimate', but 'a medium between the two'. The implication was that as an instructor of these middle rank artisan learners, Holyoake considered himself to be among those of the 'ultimate' class, now qualified to address his erstwhile peers.[68]

Such distancing tropes may have made Holyoake sound more like a middle-class educationist than a self-improving artisan. Edward Royle explained his apparently condescending pedantry in terms of felt class disadvantages; autodidactic artisans with aspirations to be accepted by their 'betters' necessarily became snobbish.[69] Alternatively, Holyoake's increasing concern with propriety in speech and writing has been understood in terms of an acutely felt need to bridge a widening social gulf between working- and middle-class radicals noticeable by the late 1840s. Due at least in part to the deepening economic disparity as well as the removal of legal liabilities for dissenters that allowed middle-class radicals to become more vocal within established society, the gulf may have been reflected in the more discernable differences in speech and language used by the groups. Artisan and working-class freethinkers sought to imitate the increasing formality of establishment prose. The attempt itself may have served to indicate the degree to which the classes and their respective language characteristics were actually drifting apart, with the unintended effect of making the groups more conscious of their differences.[70]

These explanations still leave the question unanswered: why did Holyoake seek to breach the social gulf in the first place – that is, unless, with Edward Royle, we take Holyoake to be a mere a social climber? One problem with the latter interpretation is that it fails to account for Holyoake's earlier freethinking career. Surely there were other, far more propitious paths to sought-after respectability than imprisonment for blasphemy. However, in making freethought respectable, Holyoake may have sought to transform it into a vehicle for social mobility, his own included. Nevertheless, this reading of Holyoake's career path lacks symmetry; working-class intellectuals are denied the legitimate intellectual and political motives afforded their social superiors. We must assume that Holyoake lacked integrity, that his motives were purely self-serving.

The answer may lay somewhere between Royle's interpretation and Holyoake's own assertions, and may be explicable in terms of his cross-loyalties, which were necessary due to the objectives of Secularism. Or perhaps we might say that Secularism developed as it did, in lieu of another kind of movement, due to Holyoake's cross-loyalties.[71] In any case, Holyoake saw his prospective readers and supporters as coming from different social classes and educational attainments. His primary allegiance may have been to his radical artisan roots, and working-class causes and associates, but he paid additional homage to middle-class radicals whom he considered well disposed to help his associates and their causes. With the *Reasoner*, he attempted to mediate and negotiate broader socio-political interests in an attempt to forge a motivated readership from a widened social base, whether considered in terms of class or religious categories. Holyoake believed that addressing cultural differences was a prerequisite for minimizing economic disparity. In part, Secularism was founded as a form of cultural mediation to do just that.[72]

Perhaps due to the relative success of the *Reasoner* in securing both the praise and monetary support of 'the comparatively munificent subscribers – whose bequests have exceeded my expectations as has the honour of approval from such quarters',[73] Holyoake began to identify more closely with such supporters, and to further his attempts to appease them. While in 1847, the editor was yet 'proud to address the class of "artizans"', he boasted 'that more than a third of the *Reasoner's* reporters are of the middle and educated class, quite familiar with the higher forms of reasoning'.[74] But the periodical still bore marks of Holyoake's own background, apparent in the resentment buried within a sardonic commentary regarding the advantaged scholar:

> When I contemplate the appliances which learning and science present to the scholar, and see how multiplied are his means of knowing the truth upon all subjects, I cannot conceive that he can be struggling like the untaught thinker between right and wrong. To the scholar, truth and falsehood must be apparent; and since the learned do not penetrate to the intellect of the populace, and establish intelligence among them, it must be that the learned want courage or condescension, or that common sense among them is petrified in formulas. We want either a hammer or a fire, to break the spell or dissolve the ice.[75]

From the radical artisan standpoint, the difference between 'right and wrong', 'truth and falsehood', was precisely what most of the advantaged scholars apparently either had no knowledge of or had no concern for. Otherwise they would have written and acted on behalf of the greater part of humanity to improve their lot. Granting this knowledge for the moment, however, they must therefore 'want courage or condescension', or else 'common sense among them is petrified in formulas'.[76] Whatever the reason for their failure to propagate the 'truth' – and Holyoake never seemed to consider the possibility that their class interests precluded it – the failure was motivation for Holyoake, who descried in it his own role. Such a lack defined the need for the artisan intellectual, whose new role was to mediate between the intellectuals of both classes in order to both inspire a knowledge of political right and wrong and to establish a circuit for lifting up the greater mass of mankind into the light of knowledge, which was impossible 'till subsistence is secure and leisure abundant'.[77] In these moral, economic, and educative functions, Secularism was the primary heir of Owenism's unique contribution to radical artisan politics. Yet unlike Owen's paternalism, given the goals set out in the *Reasoner*, freethinking artisans saw education as a two-way rather than strictly top-down matter.[78] Holyoake later asserted that the views expressed in the *Reasoner* 'were widely accepted by liberal thinkers of the day, as an improvement and extension of free thought advocacy', demonstrating at least his intention that the *Reasoner* educate and influence his social superiors.[79]

In July of 1849, Holyoake initiated his foray into radical middle-class literary circles with a review of George Henry Lewes's *The Life of Maximilien Robespierre* in the *Reasoner*.[80] He sent a copy of the review along with other numbers of the periodical to the biography's author at Bedford Place. Although unsure how long the papers had 'been lying there' before taking notice, by August, Lewes had read the review and was impressed with its 'tone & talent' although 'dissent[ing] from most of its conclusions'. In the company of Thornton Hunt, the son of radical poet Leigh Hunt, Lewes fired off a missive to the *Reasoner* offices and invited Holyoake for a cigar the following Monday, a night that Hunt was also available.[81] Thus began lasting friendships that signaled Holyoake's most significant literary success and began the bridge building to respectable society that would gain him admittance into the salons of numerous literary, political, and scientific luminaries of the day. The connections initiated the cross-pollination of working- and middle-class freethought that resulted in the development of Secularism proper. Doubtless, Holyoake's notoriety as a leading artisan radical and journalist, who was yet still safe to associate with – at this point presumably the last to serve jail time for atheism[82] – had facilitated this welcome into this middle-class radical society, where he met and discussed politics and philosophy with the legatees of philosophical radicalism, including Francis Place, Robert Owen, W. H. Ashurst, Francis W. Newman (see Chapter 4), Thornton Hunt, George Henry Lewes, Harriet Martineau, Herbert Spencer, Louis Blanc, and others.[83] A few of these heterodox thinkers would even contribute articles to the *Reasoner*.

As a liberal activist, rising journalist and son of the heterodox poet Leigh Hunt, Thornton Hunt was a gentlemanly counterpart of Holyoake. The two became fast friends despite Holyoake's humbler background and Hunt's open affair with Lewes's wife, Agnes.[84] Such libertinism if undertaken by a working-class radical like Holyoake would have been a greater scandal. By the end of 1849, Hunt already considered Holyoake an intimate to be included in his various activist schemes. His organizational plans for a 'Confidential Combination' of freethinkers and a 'Political Exchange' may have proven significant for Secularism. Edward Royle considers the Political Exchange foundational.[85] But the draft proposals that Hunt sent to Holyoake suggest that the Confidential Combination, with which the former has been confused, was envisioned as a means to enlist wary middle-class freethinkers into an anonymous group where they might voice advanced opinions on 'politics, sociology, or religion' without fear of reprisal.[86] The Political Exchange, on the other hand, never came to fruition, and Hunt's proposal makes clear that it was intended as a public group for the comingling of persons of various political persuasions, not as an organization for the advancement of radical thought.[87] Considering Hunt's confessions to Holyoake in correspondence regarding his position on marital relations and his lack of respect for 'the existing moral code in this country',[88] one may surmise that the 'sociology' to be discussed at the Confidential Combination had at least something to do with marital policy and a scientific system of morality, and 'religion' with secular ideas, both of which might involve 'opinions considerably in advance of those which they [publicly] avow'.[89] The club's purpose was to circumvent '[t]he tyranny which keeps down the expression of opinion in our time, [which] though less dangerous than it has been in times past, is more domesticated, more searching, and constraining'.[90] This anonymous club no doubt included Holyoake, Lewes, Hunt, Herbert Spencer, W. Savage Landor, W. J. Linton, W. E. Forster, T. Ballantine, and George Hooper, all of whom became contributors to the *Leader*. The *Leader* soon began to frame its goals in terms of 'the New Reformation', which closely resembled the goals of Secularism. Francis W. Newman, whose book *The Soul, its Sorrows and Aspirations* (1849) greatly impressed Holyoake (see Chapter 5), was among those, including Hunt and the pantheist William Maccall, who encouraged the formation of a club.[91] The members of the Confidential Combination met at the Whittington Club at the old 'Crown and Anchor' on the Strand. There they must have discussed Secularism, the New Reformation, and 'the Church of the Future', the latter phrase soon to be part of the subtitle of a Newman book published in 1854.[92] Holyoake regularly conversed with Herbert Spencer, whom Holyoake described as having 'a half-rustic look' and 'gave the impression of being a young country gentleman of the sporting farmer type'.[93] Spencer and Holyoake remained life-long friends, with regular correspondence continuing to 1894.[94]

Another overlapping milieu included the Muswell Hill circle, based in the Ashurst family home, which was also a center for radicalism and republicanism – notably in support of Giuseppe Mazzini.[95] W. H. Ashurst, '[Robert] Owen's lawyer and advisor to a generation of radical leaders', encouraged Holyoake in the development of the new Secularist movement and with one hundred pounds bankrolled the reissue of the *Reasoner* in 1849.[96] It was to Ashurst, writing to the *Reasoner* under the pseudonym 'Edward Search', that Holyoake owed the use of the words 'Secular' and 'Secularist' to describe the new branch of freethought then under formation. Holyoake added the word 'Secularism' to describe 'the work we have always had in hand'.[97] The anonymous club was undoubtedly a breeding ground of middle-class support for the budding Secularist movement and served to germinate the program of Secularism eventually expounded by Holyoake.[98]

Hunt's aspirations for the public voicing of radical opinion was more nearly realized with the weekly newspaper, the *Leader*, founded in 1850 and edited by himself and Lewes. In March 1850, Hunt sent the prospectus for the periodical to his friends, including Holyoake, and the paper began publication on the thirtieth. The weekly positioned itself at the forefront of liberal opinion. George Lewes was responsible for the reviews of literature and the arts and Marian Evans assisted him with editing and writing. Hunt was the chief political editor and contributor. Holyoake had secured the premises in Crane Court, was retained as the business manager, and contributed regular articles on the cooperative movement under the pseudonym, 'Ion'.[99]

Many from this same circle of London writers also met at 142 Strand, the home and publishing house of John Chapman, the publisher of the *Westminster Review*, the organ of philosophical radicalism.[100] Contributors to the periodical included Lewes, Marian Evans (soon to adopt the penname of George Eliot), Herbert Spencer, Harriet Martineau, Francis W. Newman (see Chapter 5), Charles Bray, George Combe, and, by 1853, Thomas Huxley. Chapman published a series of books heralding a new religion they entitled the 'New Reformation', including Francis Newman's *The Soul* (1849) and *Phases of Faith* (1850); Robert William Mackay's *The Progress of the Intellect* (1850); Herbert Spencer's *Social Statics* (1851); and Leigh Hunt's *The Religion of the Heart* (1853). (See Chapter 5.)

Many of the *Westminster* writers also showed an interest in the writings of Auguste Comte 'and in his platform for social improvement through a progressive elaboration of the sciences'.[101] Marian Evans reviewed for the *Westminster* Mackay's *The Progress of the Intellect*, a work of Comtean orientation.[102] Holyoake came to know Comte's ideas through his association with Lewes and Evans, as well through Harriet Martineau, who was then preparing her translation of his *Positive Philosophy*. Holyoake advised Martineau to help her find a translator for mathematical parts of the *Positive Philosophy*.[103] In 1855, Holyoake went so far as to visit Comte in his apartment in Paris, expressing an interest in publishing an English translation of the third volume of the *Système de Politique Positive* (1851–1854), which Holyoake would entitle *Philosophy of History*. Comte approved of the project and, famously naïve, even entertained hopes of an English popular uprising that would end with Holyoake installed as a dictator, so that Holyoake could order society according to positivist principles. Yet Comte's *Appeal to Conservatives* (1855) may have worried Holyoake given its appeal to the right, as Holyoake did not reply to the book or a letter Comte had sent him. When Comte asked his British devotee Richard Congreve to write to Holyoake to ask about the translation of the third volume of the *Système*, Holyoake responded that he was still interested in the project, but had not yet started. Yet John Fisher, another of Comte's British acolytes, reported to Comte that Holyoake was already publishing extracts from the *Système* in weekly installments in the *Reasoner*. In fact, by this time, the subtitle of the *Reasoner* was *'Journal of Freethought and Positive Philosophy'*. Although Comte criticized Holyoake as an agitator, he nevertheless entertained hopes that Holyoake's excerpts would gain some new adherents for positivism.[104] Meanwhile, Holyoake's contact with Comtean ideas was essential for the step that he was contemplating – to take freethought in a new direction.[105] In the *Reasoner* in the 1850s, Holyoake regularly cited Comte's famous phrase, 'Nothing is destroyed until it is replaced', which he appropriated for Secularism.[106] Like Comte, Holyoake believed that religion had to be replaced with a 'positive' creed rather than being simply negated by atheism.

Martineau approvingly noticed the new direction that Holyoake was taking freethought:

> The adoption of the term Secularism is justified by its including a large number of persons who are not Atheists, and uniting them for action which has Secularism for its object, and not Atheism ... [I]f by the adoption of a new term, a vast amount of impediment from prejudice is got rid of, the use of the term Secularism is found advantageous.[107]

The *Westminster Review* ran an article on Secularism in 1853, stressing that with Secularism, freethought had 'abandoned the disproof of deity, contenting itself with the assertion that nothing could be known on the subject'.[108] In 1862, the *Westminster* claimed, rather wishfully, that Secularism had become the belief system of the silent majority of the working classes, whatever the number of those who subscribed to its periodicals or associated with its official organizational structures.[109] Here, the author echoed the earlier remarks about Secularism by Horace Mann in his Introduction in 1854 to the 1851 census on religious worship, although without Mann's histrionics:

> There is a sect, originated recently, adherents to a system called 'Secularism'; the principal tenet being that, as the fact of a future life is (in their view) at all events susceptible of *some* degree of doubt, while the fact and the necessities of a present life are matters of direct sensation, it is therefore prudent to attend exclusively to the concerns of that existence which is certain and immediate – not wasting energies required for present duties by a preparation for remote, and merely possible, contingencies. This is the creed which probably with most exactness indicates the faith which, virtually though not professedly, is entertained by the masses of our working population; by the skilled and unskilled labourer alike – by hosts of minor shopkeepers and Sunday traders – and by miserable denizens of courts and crowded alleys. They are *unconscious Secularists* – engrossed by the demands, the trials, or the pleasures of the passing hour, and ignorant or careless of a future. These are never or but seldom seen in our religious congregations; and the melancholy fact is thus impressed upon our notice that the classes which are most in need of the restraints and consolations of religion are the classes which are most without them.[110]

In short, Holyoake's role in the middle-class London literary and intellectual *avant garde* meant that he had moved from the radical artisan fringes to become a central figure in London radical circles; his '"Secularism" was their watchword', and the *Reasoner* the leading propagandist organ. At age twenty-five and not yet a fellow in the Royal Society, Huxley was introduced to the leading lights in the scene, including Spencer, Lewes, Marian Evans and, undoubtedly, Holyoake.[111] As a writer for the *Westminster* by 1853, he could not have but taken notice of the new notion of Secularism then in circulation.

By the early 1850s, the cross-pollination between the middle- and working-class freethought movements was well underway. Holyoake's reviews and notices of the works of Francis Newman, Lewes, Martineau and others in the *Reasoner*, together with his work at the *Leader* and the notices of his Secularism in the *Westminster*, completed a two-way circuit of exchange. Holyoake's alliance of artisan and middle-class advocates preceded by over thirty years the more successful attempt by the son of the famous Secularist Charles Watts, Charles Albert Watts, who appropriated the idea of agnosticism for his *Agnostic Annual* in 1884, 'to move towards an alliance with eminent middle-class unbelievers and away from secularism's radical working-class roots'.[112] Secularism, while never disavowing its class roots, had by mid-century already forged alliances with eminent middle-class unbelievers and liberal theists, who were attracted to the new movement's program of greater inclusion.

Holyoake was admittedly flattered by his reception among middle-class intellectual circles, and boasted of it in his writing. He paid tribute to Eliot and Lewes in his book *Bygones Worth Remembering* (1905), stating that until he had been accepted into such company his had been 'an outcast name, both in law and literature'. His inclusion in the *Leader* was 'the first recognition of the kind I have received'.[113] But this conciliation with non-atheists and middlebrow radicals was seen by many of Holyoake's older working-class acquaintances as the gentrification of working-class infidelity as it merged with the gradualist, middle-class scientific meliorism ascribed to George Eliot by Charles Bray and others:

> She held as a solemn conviction ... that in proportion as the thoughts of men and women are removed from the earth ... are diverted from their own mutual relations and responsibilities, of which they alone know anything, to an invisible world, which alone can be apprehended by belief, they are led to neglect their duty to each other, to squander their strength in vain speculations ... which diminish their capacity for strenuous and worthy action, during a span of life, brief indeed, but whose consequences will extend to remote posterity.[114]

This view was representative of Secularism, which evolved philosophically in connection with such middle-class influences and was developed by Holyoake expressly in order to accommodate them. Middle-class unbelief had benefitted legally and ideologically from the artisan- and working-class freethought movement, which under Holyoake parted ways with radical working-class politics as the latter tended toward the negative secularism of Charles Bradlaugh on the one hand, and eventually toward Marxist socialism on the other. Meanwhile, middle-class skeptics legitimated Holyoake's brand of Secularism.

Atheism, sex, and secularism

On 5 April 1877, as was widely reported in the press, Annie Besant and Charles Bradlaugh were arrested and charged with printing and publishing 'a certain indecent, lewd, filthy, bawdy, and obscene book, called "Fruits of Philosophy," thereby contaminating, vitiating, and corrupting morals'.[115] Besant and Bradlaugh would stand trial for the publication, a trial that would gain enormous publicity and bring significant, and for some, unwanted attention to the Secularist movement. For Besant and Bradlaugh, the Knowlton affair, as it came to be called, represented a test of a free press, as well as the defense of 'a discussion of the most important social question which can influence a nation's welfare'.[116] This discussion involved the doctrine of population and the right of a free people to critically examine the issue of birth control. Although the trial ended in February 1878 in an acquittal on the grounds of a technicality exploited by Bradlaugh, the savvy former legal clerk, the trial put contraception 'onto the breakfast tables' of the middle class and associated it with Secularism.

As I discuss below, *Fruits of Philosophy* had been one in a line of neo-Malthusian pamphlets published by the freethought movement since the days of Richard Carlile. But the republication of the manual in March 1877 by Besant and Bradlaugh would become a major source of controversy and serve to roil the Secularist movement, and, according to some, split the Secularism movement as it never had been divided before. As I show, however, the republication of *Fruits of Philosophy* and the subsequent trial for obscenity only served to calcify a long-standing rift within the Secularist leadership and ranks.

Dr. Charles Knowlton wrote and first published *Fruits of Philosophy, or the Private Companion of Young Married People* in 1832 in Massachusetts. The pamphlet was a neo-Malthusian pro-birth-control manual detailing the physiology of human sexuality and the means of couples for limiting the size of their families. In the 'Philosophical Proem' introducing the text, Knowlton argued that the practice of sex was a physiological and moral necessity; he reasoned from Benthamite principles that any moderate expression of sexual passion that did not result in misery added a net pleasure to the world and thus was to be encouraged. Furthermore, the sexual instinct would not be curbed in the mass of humanity according to Malthusian abstentionism. Only practical measures to limit procreation – new methods of contraception – could thus solve the predicament resultant from the sexual instinct on the one hand and the tendency of population growth on the other.[117] Although the pamphlet was released anonymously, Knowlton was arrested, tried, and convicted of obscenity, serving three months of hard labor in East Cambridge jail.

Fruits of Philosophy was imported into Britain and published by the radical disciple of Richard Carlile, James Watson, who took over Carlile's publishing ventures while Carlile was in Dorchester jail. Watson also became Holyoake's publisher and in 1853 Holyoake bought Watson's stock and sold it under the Secularist banner. As noted by Bradlaugh and Besant in their chronicling of the Knowlton affair in the Publisher's Preface of their republication of the work, *Fruits of Philosophy* was listed in Holyoake's 'Freethought Directory' in 1853.[118] The *Reasoner* had sometimes listed the birth control pamphlet among the books sold by Holyoake's Fleet Street House for Watson (although Holyoake had never explicitly supported the publication).[119] *Fruits of Philosophy* was published for a time by Austin Holyoake, George Holyoake's brother, in conjunction with the *National Reformer*, and when Watson died, the plates for all of his publications, including *Fruits of Philosophy*, were purchased from Watson's widow by Charles Watts, who published the work until 23 December 1876.[120]

As a publisher of *Fruits of Philosophy*, it was Watts who, in January 1877, was first charged with printing and publishing an obscene book. The legal attention attracted by the work was probably due to several factors, not the least of which included new drawings inserted by Watts, and his lowering of the price.[121] But another factor was the passage in August 1857 of the Obscene Publications Act, which made a court's interpretation the new test for obscenity. According to the new Act, a publication could be deemed obscene if it demonstrated – as argued successfully by Lord Chief Justice, Sir Alexander Cockburn in 1868 in the celebrated case of *Regina v. Hicklin* – a 'tendency ... to deprave and corrupt those whose minds are open to such immoral influences, and into whose hands a publication of this sort may fall'.[122] Obscenity, that is, was now legally in the eye of the beholder, rather than based on something 'objective' in the text itself. The law apparently emboldened prosecutors and facilitated arrests. Further, given this new definition of obscenity, the accused was effectively guilty until proven innocent.[123]

After his arrest, Watts met with Bradlaugh and Besant, who agreed to support him in his defense and to raise money for his trial. But upon further reflection, once out of Besant's and Bradlaugh's company, Watts decided not to defend the right to publish the book and to recant his not-guilty plea and enter a plea of guilty as charged. Upon his trial,

Watts was fined 500 pounds and released.[124] Besant and Bradlaugh not only immediately cut their business ties with Watts, who had been their publisher for the *National Review* and other works but also they decided to republish *Fruits of Philosophy* under the banner of their newly formed publishing partnership, the Freethought Publishing Company.[125] While they found much wanting in *Fruits of Philosophy*, the right of publication, they argued, was a matter of principle. Bradlaugh and Besant reasoned that if they failed to assert 'The Right of Publication' of a book that was not obscene but was also a scientific text, then the freethought movement would be damaged and the cause of a free press severely compromised.[126]

Not everyone in the Secularist movement agreed with this decision to republish, least especially Holyoake, who (unsuccessfully) attempted to remove Bradlaugh and Besant from the Executive of the National Secular Society (NSS).[127] In 1877, in the midst of the Knowlton affair, Holyoake was invited by freethinkers to chair a committee charged with reviewing the rules of the NSS. The commission challenged the position of president itself, a position that Bradlaugh had held from the beginning of the organization. The failure to rid the NSS of the presidency and thus to unseat Bradlaugh led to the formation of the British Secular Union (BSU) in August 1877, a new organization of Secularism established in opposition to the Bradlaughian NSS and supported by the new periodical the *Secular Review* as its official publication.[128] This organization, I suggest, was the result of more than the Knowlton affair; it registered a long-standing alienation between Holyoake and Bradlaugh, and their respective camps. But the secession of George Holyoake, Charles Watts, and other Secularists from the NSS, and their founding of the BSU in the wake of the Knowlton affair, solidified an already significant breach within the Secularist movement, one that now appeared to ossify around the issue of sexuality.

In his study of Darwin and respectability, Gowan Dawson devotes a chapter to obscenity legislation in connection with Darwinism, treating in some detail the relationship between the Darwinian scientific naturalists and the two branches of freethought, which Michael Mason has referred to as the 'anti-sensual progressive' (Holyoake) and the 'pro-sensual' (Bradlaugh) Secularist camps.[129] Dawson suggests that the primary division between the Secularist camps was predicated on differences over sexual policy and birth control. According to Dawson, Bradlaugh and Annie Besant's republication and legal defense in 1877–1878 of Knowlton's *Fruits of Philosophy* became the primary reason for the split between the Holyoake and Bradlaugh camps. Birth control and sexual policy, Dawson argues, 'were by far the most divisive issue[s] within the British freethought movement in the nineteenth century'.[130]

In figuring sexual policy as the fault line dividing the two Secularist camps, Dawson overlooks the well-documented, fundamental division within Secularism. This division, as Royle points out, not only took hold between the major two camps of Secularism, but also within them.[131] The primary split dated to the early 1850s and went to the definition of Secularism itself. Differences in sexual policy may be in large part understood in terms of this fundamental split. From the beginning of the movement and creed, Holyoake had differentiated Secularism from the older freethought movement, shifting its emphasis from a 'negative' to a 'positive' orientation. Philosophically, this entailed what he and others sometimes called a 'suspensive scepticism',[132] which included not only denying atheism as a requisite commitment but also definitively disavowing any declarative assertion on the question of deity. As Holyoake argued (rather misleadingly) in the celebrated debate with the Reverend Brewin Grant in 1853, '[w]e have always held that the existence of Deity is "past finding out," and we have held that the time employed upon the investigation might be more profitably devoted to the study of humanity'.[133] In terms of strategy, as we have seen, this position meant cooperation between unbelievers and believers; the invitation to join the Secularists extended not only to Christian Socialists such as Charles Kingsley and his ilk but also to liberal theists with reformist politics, such as Francis M. Newman and James Anthony Froude. In terms of principle, it meant that Holyoake's Secularism, as opposed to Bradlaugh's, was specifically not atheist.

Many leading freethinkers rejected the construction that Holyoake had put on freethought with his Secularism, however, as well as his aversion to centralized organization and purported failures in organization. These included, as we have seen, Charles Southwell; but the defectors also included Holyoake's brother Austin, Robert Cooper, and most importantly, Charles Bradlaugh. Cooper started the Secularist *Investigator* in 1854 explicitly to challenge Holyoake's *Reasoner*. Although it operated out of Holyoake's 147 Fleet Street publishing and propaganda headquarters, was published by Holyoake, and adopted the moniker of Secularism in its subtitle, Cooper's periodical granted Holyoake no exemption from fierce criticism. Cooper established the *Investigator* to counter to what he deemed Holyoake's conciliatory approach. In its opening salvo, the statement of policy, Cooper mocked Holyoake's *Reasoner*, Holyoake's association with the *Leader*, as well as his sanctioning of the theology of feeling expounded by Francis Newman, Leigh Hunt, and others (see Chapter 5):

> The age of expediency in theology, as in politics, is past. We aim to be a Reasoner, but not a Trimmer. Aspiring enough we may be as a Leader, but not weak enough to be a Feeler. A true Reasoner is one who has the courage to follow reason *wherever* it leads him. A legitimate Leader is one who is in *advance* of his contemporaries, labours to develop the full object of his mission, and bring the world to it.[134]

Holyoake's position was taken as accommodation adopted out of weakness and timidity, rather than philosophical conviction and sound policy. Secularism's 'positive' approach evinced a lack of courage in the face of an overwhelming enemy. The importance granted to feeling for the moral sense was best reserved for sentimental religionists.[135]

The *Investigator* also published the opinions of Southwell, who belittled the new direction that Holyoake was taking freethought:

> Superstitionists are laughing at us; and well they may. Their worst enemies in name are their best friends, in fact; the impersonal policy, of which we hear so much, is a hoax, by which they, and they only, profit. It spares the feelings of sacerdotal knaves, while outraging the feelings of those who groan under the heavy pressure of their frauds, as I proved in my recent but unpublished discussion with Mr. Holyoake.[136]

With Bradlaugh's meteoric rise to prominence in the Secular field in the 1860s, the divide between the Secularist camps became more pronounced. In 1850, Holyoake had chaired a freethought meeting and invited the young Bradlaugh, at the mere age of seventeen, to speak on 'The Past, Present, and Future of Theology'.[137] By the late 1850s, Bradlaugh had found in the *Investigator* a vehicle for his trenchant atheism. In 1858, he had been elected president of the London Central Secularist Society, assuming the position Holyoake had held for nearly a decade. By 1860, he had become founder and co-editor of the *National Reformer*. Yet in an attempt to close the ranks of the Secularist body, in November 1861, Bradlaugh invited Holyoake to join the *National Reformer* as a special contributor. Holyoake accepted, and even signed a letter entitled, 'One Paper and One Party', published in the periodical. Beginning in January 1862, he was responsible for curating three pages – either of his own writing, or from his associates. But in February, a correspondent to the paper complained of the paper's diversity of opinion and asked what the *National Reformer* definitively advocated regarding religion. Bradlaugh's answer effectively marked the end of Holyoake's involvement: 'Editorially, the *National Reformer*, as to religious questions, is, and always has been, as far as we are concerned, the advocate of Atheism'. The consequence was a fall-out between Bradlaugh and Holyoake that included a financial dispute, with Holyoake apparently demanding a year's salary, after having only served three months in his capacity as 'chief contributor'.[138]

By 1870, the lines were even more severely drawn. In a debate between Holyoake and Bradlaugh (chaired by Holyoake's brother, Austin, by then an acolyte of Bradlaugh's), the topic was the place of atheism within Secularism. In effect, George Holyoake denied that Bradlaugh was a Secularist at all. Further, Bradlaugh admitted that, according to Holyoake's definition (a definition, he suggested, that the founder of the movement had a right to maintain), Holyoake was right that he should not be called a Secularist.[139] Nevertheless, by then the President of the NSS, Bradlaugh asserted that Secularism necessarily amounted to atheism – 'I hold that Atheism is the logical result to all who are able to think the matter out' – and that Holyoake's reasoning was simply flawed.[140] Holyoake, for his part, remained as firm as ever that Secularism did not 'include' atheism, but concomitantly, that it did not 'exclude' atheists,[141] a point which Bradlaugh considered illogical.[142] Holyoake further suggested that making atheism a condition of Secularism was to delay the work of Secular improvement indefinitely, while atheism made its clean 'sweep' of theological notions:

> Mr. Watts [then still a Bradlaugh supporter] goes on to state [in the *National Reformer*], 'The province of Secularism is not only to enunciate positive principles, but also to break up old systems which have lost their vitality, and to refute theologies which have hitherto usurped judgment and reason'. *Here is an immense sweep*. None of us will live to see the day when the man who has made it, will be able to give us the secular information which we are waiting to receive now.[143]

Instead of advocating the undertaking of such 'an immense sweep', Holyoake contended that Secularism should be established independently of theology as a creed that had positive principles of its own and the work of secular improvement should be undertaken at once. He quoted a contributor to the *National Reformer* (again, his brother, Austin), who had asserted that it was 'impossible to advocate Secular principles apart from Atheism ... There is no man or woman who is willing to listen to Secular views, knowing they are intended to set up a system entirely apart and devoid of all religion'. George Holyoake did not spare his brother criticism:

> You set up Secular principles *for their own value*. Many persons are Secularists who can see religion even in this. *The provision is not to set up a thing 'devoid of all religion', but to set up a thing distinct in itself,* and you have no more right to say it is set up apart from the religion, than the clergyman has a right to say, when you set up Secular knowledge apart from his creed, that you intend thereby to set it up *devoid* of religion or public piety.[144]

We see here that by Secularism Holyoake meant a substantive doctrine, not the mere absence or negation of religion or religious belief. For this reason, it could (logically or otherwise) stand parallel to (or above) religious systems. Moreover, he was even willing to allow Secularism to be construed as a religion in its own right. This was a more acceptable option than including atheism as a necessary element of Secularism.

Furthermore, whenever the question of sexual policy was raised, the issue of atheism was never far removed. In the 1870 debate between Bradlaugh and Holyoake, for example, Holyoake had distinguished between what he called 'positive' and 'negative' atheism. While the former was 'a proud, honest, intrepid, self-respecting attitude of the mind', 'Negative Atheism' consisted of 'mere ignorance, of insensibility, of lust, and gluttony, and drunkenness, of egotism or vanity'.[145] With this distinction, which he registered seemingly out of the blue, Holyoake was in fact acknowledging a long-standing association of atheism with immorality, in particular with sexual profligacy and other sensual licentiousness. His definitions represented a not-so-subtle chastisement of the Bradlaugh camp for its neo-Malthusian advocacy in the *National Reformer* – its recommendations of preventive checks to procreation (birth control). Moreover, Holyoake also apparently commented on the position of his brother, Austin, whose own neo-Malthusian pamphlet, *Large or Small Families*, had appeared in 1870. While Bradlaugh denied knowledge of any such 'Negative Atheism' or anyone who practiced it,[146] given his well-known neo-Malthusianism, it must have been clear to those familiar with the contentious field of Secularism what Holyoake meant by the phrase 'Negative Atheism'.

Although Holyoake rarely had addressed the issue, since the publication in 1826 of Richard Carlile's *Every Woman's Book, or What is*

Love?, freethinkers had openly vetted the questions surrounding sexual policy, birth control, and free love.[147] Neo-Malthusian doctrine had been recommended anonymously for the working classes in broadsides circulated (if not entirely written) by Francis Place from the early 1820s. Prompted by Place's circulars and friendly philosophical coaxing, Carlile wrote two major installments on the subject of birth control: 'What is Love?' and *Every Woman's Book*. To Place's chagrin, Carlile's essay and book not only advocated artificial checks but also regular sexual intercourse, equality of the sexes in the solicitation of sexual activity, and a policy of relative 'free love' made possible only by contraception. Following these bold interventions, Robert Dale Owen, the son of Robert Owen, published his *Moral Physiology* in 1830, another pamphlet that acknowledged Malthus's warning about population growth, but repudiated the moral pessimism and prognostications of Malthus's disciples. Owen argued that deliberative methods of population control could accompany 'rational reform' (reforms for the alleviation of vice and misery) and thus could abet social and moral progress. But *Moral Physiology* was offered primarily as a corrective to what Owen deemed Carlile's physiological errors, poor taste, and lapses in judgment. Owen was concerned that, given its coarseness, apparent vulgarity, and promotion of women's sexual prerogatives, *Every Woman's Book* had been interpreted as recommending sexual libertinism, and thus excited the kind of prejudice that had been directed at Owen himself for supposedly recommending it in the United States. Carlile's manual, he considered, rather than helping to check the population and thus the size of working-class families, might be dismissed as immoral, allowing the principle of population to remain unheeded, or rejected outright. Thus he ventured to author *Moral Physiology* under his own name, hoping to sever associations of contraception and sexual immoderation.[148] It was Robert Dale Owen who encouraged Charles Knowlton to write and publish his *Fruits of Philosophy*.

In the Publisher's Preface to the 1877 edition, the edition that led to the obscenity indictments brought against Bradlaugh and Annie Besant, Bradlaugh and Besant charged Holyoake and company with hypocrisy, suggesting that he and Watson had sold and profited by the book for decades. If they had considered the book obscene all the while, then they had carelessly 'thus scattered obscenity broadcast over the land'.[149] Otherwise, why did they not stand behind the republication of the book? Holyoake's disapproval of the decision by Bradlaugh and Besant to republish and defend by the book had been registered by the time they wrote their publisher's preface, given Holyoake's disavowals in the press.[150] It was clear that Bradlaugh and Besant were already acutely aware of Holyoake's position.

Neo-Malthusian doctrine necessarily involved Secularists of the Holyoake camp in a moral quandary. Should birth control apply strictly to the moderation of family growth within the confines of marriage? If not, might it encourage sexual profligacy? Given his concern for Secularism's respectability, Holyoake had always recommended moral discipline and reservation. Although possibly having some sympathy for neo-Malthusian practices within marriage, having supported more liberal laws for divorce, and despite his contact with Hunt and Lewes, he had for decades effectively skirted the issues invoked by freethought in connection with sexual policy.[151] Further, with roots in the communitarianism of Owenite socialism, the implications of Malthusian political economy had always been unpalatable. Thus, the Knowlton affair thrust him into a confrontation he would have rather avoided. The Knowlton affair had connected Secularism with neo-Malthusianism, potentially embarrassing Holyoake, and not only for the associations with immorality that he feared. Not only did neo-Malthusian doctrine, *per se*, conflict with his socialist predilections but also the problem of sexual conduct exposed theoretical and practical contradictions within his Secularism; Secularism's refusal to place primary importance on the elimination of Christian theology and morality, its insistence on suspending judgment regarding Christian values that supposedly did not conflict with secular progress – this abdication of normativity was impossible where sexual conduct was concerned. To be strictly consistent theoretically, a Utilitarian and neo-Malthusian moral code for sexuality would have signified widespread use of contraceptives and such extensive sexual activity as afforded a net pleasurable return for all concerned, regardless of the legal status of the partners. Yet Holyoake never advocated such a position. Certainly, as Michael Mason has observed, '[t]he exalted status of rationality in the advanced thought of the eighteenth century had a lasting influence on all radical and reforming creeds in the nineteenth', including Secularism.[152] But, arguably, the utilitarianism of Holyoake's Secularism was buttressed by and dependent upon prevailing Christian values, what Mason refers to as 'classic moralism', at least where human sexuality and social reproduction were concerned. Arguably, Holyoake's position on sexuality owed less to anti-sensualist rationalism inherited from the Enlightenment than it did to the observance of Christian-based propriety. As John Stuart Mill put it to Holyoake in a letter in 1848:

> [T]he root of my difference with you is that you appear to accept the present constitution of the family & the whole of the priestly morality founded on & connected with it – which morality in my opinion thorough[ly] deserves the epithets of 'intolerant, slavish & selfish'.[153]

That is, Holyoake's Secularism had not established an entirely unalloyed social science in place or independent of religious systems. Rather, in his attempt to erect a substantive creed alongside (or above), but not necessarily in contradiction to Christianity,[154] his Secularism had implicitly assumed standards for sexual conduct having little or nothing to do with its own stated principles. In terms of secularity, this meant that Secularism never *entirely* differentiated itself from the religious sphere.

Conclusion: secularism versus the standard secularization thesis

As introduced and developed by Holyoake, mid-century Secularism appeared to solve many of the problems posed by and for freethought radicalism itself, such as the *desideratum* to conduct free and open inquiry and expression without abdication to religious authority and unhampered by the legal and customary threats encountered in a social order with an official state religion. Holyoake modified freethought by pruning its atheistic rhetoric, allowing free thinkers to avoid questions regarding the supernatural and to disavow its clergy in matters relating to knowledge and morals, without the expected bombast and negation. By excluding questions of belief from morality, Secularism opened up a space where working-class and genteel radicals, atheists, theists, and 'agnostics' could potentially cooperate for the material improvement of humanity, especially the working classes.

Many freethinkers, both those of his own generation, and those to follow, differed with Holyoake's conception of Secularism, and either rejected it outright, or modified it for their own purposes. As I have suggested, the major division between the Holyoake and Bradlaugh camps was based primarily on the question of atheism but also included differences over Malthusian political economy and a pro-birth control sexual policy derived from it. But sexual policy and atheism were not so easily disentangled, such that the mere mention of one often implied the other. Finally, sexual policy represented a contradiction within Holyoake's Secularism and thus the extent to which it had failed to establish a secular system as fully differentiated from the religious sphere.

Remarkably, the two different senses of Secularism that I have discussed, at least where the primary distinction is concerned, survive to this day in the forms and understandings of general modern secularism. (And, so does confusion between them.) Under Bradlaugh's model, arguably the received contemporary understanding of secularism, the mission of secularism is evacuative, the category of the secular is negative, and secularization is understood as progressive and teleological. Secularism amounts to a gradual but eventual emptying of religion from the public (and private) sphere. That is, Bradlaugh's Secularism amounted to a belief in what we now understand as the standard secularization thesis.[155] On the other hand, under Holyoake's model, Secularism is constructive, the category of the secular is positive and substantive, and secularization is understood as an increasingly developing, complex plurality of belief, unbelief, and suspension between the two, along with other creedal commitments. As we have seen, Holyoake represented Secularism as a pluralistic, inclusive, and contingently constructed combination of willing theists, unbelievers, and, anachronistically speaking, 'agnostics'. He did this by positing improvement in this life as a common aim of believers and unbelievers, leaving metaphysical questions largely out of consideration. In this, I argue, Holyoake tacitly acknowledged the unlikelihood that Enlightenment rationality, extended into the nineteenth century, would utterly eradicate religious belief. As he put it in the 1870 debate with Bradlaugh, the complete evacuation of religiosity would require such 'an immense sweep' that to attempt it was tantamount to insanity and resulted in the gross negligence of pressing secular matters. Within a state that had recently criminally persecuted blasphemy, that is, Holyoake nevertheless grasped a sense of secularity as involving recognition and cooperation between religion and its others, a vision of the public and political spheres not unlike that which Jürgen Habermas has recently described as 'post-secular'.[156] Rather than (or even while) expecting its disappearance according to a model of secularization (or Secularism), that is, the secularist should accommodate religious discourse within a public sphere notable for its uneven and forever incomplete secularization. In fact, secularization and Secularism represented just this incomplete and permanent unevenness. And, as the question of sexual policy illustrated, Holyoake's Secularism itself was never fully secularized.

Once freethought entered this 'positive' phase, however – one of positing a substantive moral and epistemological value system, as opposed to merely antagonizing religious believers and negating theism – it could develop into a new, more inclusive, sophisticated creed and movement. Edward Royle and others have suggested that this development should be understood in terms of a kind of limited ecumenism, as the transformation of a religious sect into a denomination.[157] However, such an interpretation fails to grasp the secular as a category distinct from and yet necessarily related to and dependent upon the religious. With

Holyoake's Secularism, freethought was not, or no longer, an entirely religious movement *per se*. Instead, by virtue of a demarcation principle that removed from consideration Christianity's metaphysical convictions, the secular began a process of differentiation from within the religious sphere. With Secularism, freethought no longer contended for metaphysical sovereignty precisely on the grounds of theology itself. Or to put it another way, with mid-century Secularism, some freethinkers began to understand secularity differently. Rather than positing the category of the secular as merely the negation or absence of religion and belief, thus keeping it securely within the religious ambit, secularity (called Secularism by Holyoake and company) was understood and described as a distinct development, a new stage resulting in an overarching *condition* that embraced unbelief and belief, the secular and the religious, and not the negation of one by the other.

Laura Schwartz puts it thusly for the benefit of contemporary historiography:

> Once secularism is approached as a substantive rather than a negative category – as something more than simply an absence of religion – it becomes possible to see how religion may indeed play a role within a secular worldview without simply collapsing secularism into the wider category of religion.[158]

Schwartz is of course speaking to *our* understanding of secularity, invoking Taylor's rejection of and alternative to the standard secularization thesis – of secularization as continual 'subtraction' – and applying this new conception to the period. However, this understanding of secularity should not only guide our research but also should be recognized as precisely the conception that was dawning on Holyoake by the late-1840s, and what he consciously understood as developing with Secularism. This was in fact how Holyoake had envisaged Secularism proper at mid-century.

4

Secularizing Science: Secularism and the Emergence of Scientific Naturalism

The received notion regarding the relationship between science and secularism is that modern science is undoubtedly a secular and secularizing formation. As science advances, so this story goes, religion inevitably retreats and is eliminated from the domains of knowledge production, the public sphere, and even private belief. As we saw in Chapter 1, Richard Carlile's hopes for a materialist science were leveraged on such a narrative. Assuming this article of positivist faith, at least until the last quarter of the twentieth century, Whiggish historians and positivist sociologists continued to work under the assumption that in order to understand secularization, one should begin with science (among other factors) and chart its impact on the broader public sphere.[1] As I suggest in Chapter 2, this approach begs the question of just how science became secular in the first place – or moreover, how science came to be understood as secularizing *per se*. Matthew Stanley notes in a related context that scientific naturalism was not always the dominant philosophical framework for conducting science; its emergence and later prominence were by no means natural or inevitable.[2] Similarly, science was not always 'secular', and there is nothing inevitably secularizing about its growth and development. It should be quite clear that the belief in an essentially secular and secularizing science is itself in need of explanation (although this chapter does not aim at such an explanation, at least not directly). As the sociologist David Martin submits, '[w]hat matters is the *reception* of science and technology with respect to religion, not some intellectual and sometimes mythic history of the relationship generated in the academy'.[3] The reception of science and technology in connection with religion and the secular is predicated upon social and cultural factors, including the distance of church and science from political power and cultural authority. The variability of these relationships 'allows room for ... historical particularity and cultural contingency'.[4] This chapter demonstrates just such particularity and cultural contingency.

Rather than assessing the putative secularizing impact of science, in this chapter, I approach the secularity of science from the other side. I examine an avowedly secular cultural formation and its role in the secularization of science itself; in particular, I consider the creed and movement of George Holyoake's mid-nineteenth-century Secularism for its contribution to the emergence of a nearly coterminous scientific naturalism. However, by 'secularization', I do not mean to suggest the ultimate extirpation of religion or belief. Rather, I define secularization as the arrival at the condition of secularity: at the coexistence and co-constitution of belief and unbelief, religion and the secular, under a common umbrella. This chapter argues that Secularism as defined and developed by Holyoake was instrumental in advancing scientific naturalism as the expression of secularity within the science of the second half of the nineteenth century.

As Bernard Lightman points out in the final essay of a recent anthology, the term 'scientific naturalism' was marshaled by Thomas H. Huxley to refer to the movement in the last half of the nineteenth century to 'redefine science and transform British society'.[5] Scientific naturalism was premised on the uniformity of nature, a strict adherence to empirical findings, and an evolutionary view of nature and the cosmos. Scientific naturalism emerged as a dominant cosmological and epistemological framework for science, becoming, at least by 1870, a shibboleth for scientific validity, as well as for the exercise of cultural authority in the name of science in the public sphere. Historically, scientific naturalism is most remarkable for the cultural work, polemical support, and philosophical defense that it performed on behalf of Darwinian evolutionary theory. Its most important proponents were Thomas H. Huxley, John Tyndall, and Herbert Spencer.

Scientific naturalism and its polemical engagements with competing cosmological and epistemological frameworks has been a subject of considerable inquiry over the past forty-plus years in the historiography of Victorian science.[6] However, as Frank M. Turner pointed out over twenty years ago in *Contesting Cultural Authority* (1993):

> Although a considerable literature has accumulated about the scientific publicists [Thomas Huxley, John Tyndall, Herbert Spencer, et al.] and their polemical careers, historians have too rarely sought to understand from what previous intellectual tradition or traditions they emerged.[7]

To locate these traditions, Turner harked back to eighteenth-century rationalism and Kantian metaphysics. For contemporary sources, he pointed to the institutions and publications of the nineteenth-century popular enlightenment – 'the Mechanics' Institutes, the Owenite Halls of Science, and the publications of Knight and Chambers'. Given its artisan provenance, Turner discounted Secularism out-of-hand, because, he remarked, the new publicists 'had hoped to recruit support from the upper and middle classes'.[8] According to this interpretation, Huxley and fellow scientific naturalists ignored Secularism because they wanted to avoid the taint of lower-class infidelity that clung to it. With this dismissal, however, Turner not only overlooked the fact that Secularism can be tracked to the popular enlightenment and Owenism, among other sources, but also that Holyoake's Secularism had succeeded in securing middle-class support, including that of the scientific naturalists themselves. Finally, Secularism's importance for scientific naturalism does not depend on the public acknowledgment of the same by the scientific naturalists. It is certainly possible that Secularism was an important source for scientific naturalism without being credited as such. In fact, Secularism's supposed infidel taint helps explain why the scientific naturalists rarely credited it publicly it as a source.

Much work has been done over the past twenty-plus years to roll back the advent of scientific naturalism from the watershed publication event of *The Origin of Species* in 1859 to earlier decades.[9] Historians have noted that freethinking radicals from the 1840s (notably, those connected with Holyoake) drew upon pre-Darwinian evolutionary doctrines to support their social and political objectives; Adrian Desmond has shown that, even before the publication of *Vestiges of the Natural History of Creation* (1844), radical artisan and working-class science advocates marshaled doctrines of species transmutation to advance their anti-clerical, democratic and leveling socio-political programs.[10] More recently, John Van Wyhe has argued for the importance of phrenology to the emergence of scientific naturalism by mid-century. Drawing on Robert M. Young's metaphor, 'the river of nineteenth-century naturalism was fed by many streams', Van Wyhe argues that phrenology represented 'another important fountainhead of naturalism'.[11] Still others have focused on the promotion of scientific naturalism by *later* Secularists. In particular, Bernard Lightman has examined the late Victorian appropriators of agnosticism for their role in the spread of scientific naturalism to cloth-cap readers, and Suzanne Paylor has studied the role of Edward B. Aveling in popularizing Darwinian evolution and in turn bolstering atheism.[12] Despite the search for precursors and

Secularism's cultural and historical contiguity and philosophical family resemblance to scientific naturalism, Holyoake's Secularism has yet to receive the full credit it deserves as a cultural and intellectual forebear of scientific naturalism.

Extending and revising the findings of an earlier study,[13] I argue here that Secularism was a significant source for what James R. Moore referred to as the new 'creed' of scientific naturalism to emerge from the mid-nineteenth century.[14] Not only did Holyoake's Secularism help clear the way for the scientific naturalists by fighting battles with the state and religious interlocutors but also it served as a model for what Huxley, almost twenty years later, termed 'agnosticism', an earlier coinage than 'scientific naturalism', and one intended to set the limits for scientific knowledge,[15] while arguably serving to protect the scientific naturalists from charges of infidelity and atheism. Further, the term 'scientific naturalism' was first used positively to describe the epistemology of the naturalists in the pages of a periodical deriving directly from Holyoake's Secular lineage.

Holyoake modified freethought in late 1840s and early 1850s, as he forged connections with middle-class literary radicals and budding scientific naturalists, including Benthamite utilitarians, liberal theists and religious skeptics, some of who met in a 'Confidential Combination' of freethinkers. Secularism became the new code word for this coterie. (See Chapter 3.) Later, as I show below, Secularism promoted and received reciprocal support from the most prominent group of scientific naturalists, as Holyoake used Bradlaugh's atheism and neo-Malthusianism as a foil, and maintained relations with Huxley, Spencer and Tyndall through the end of the century.[16] The circuit of exchange that I trace between Holyoake's camp and the scientific naturalists suggests that Secularism had been important to scientific naturalism all along – well before the scientific naturalists incorporated Darwinism into their program – offering a form of naturalism from which Huxley could borrow, and softening the religious animus against naturalistic forms of thought. Although Secularists 'made little use of Darwin' before Charles Albert Watts's emphasis on evolutionary theory in the early 1880s,[17] in Holyoake's Secularism, we find the beginnings of the mutation of radical infidelity into the respectability necessary for the acceptance of scientific naturalism as well as the distancing of later forms of freethought incompatible with it. Holyoake's Secularism thus represents an important early *stage* of scientific naturalism. Further, the importance of Secularism – a cultural, political and social movement and philosophical creed – underscores the historical contingency of the secular in connection to science. Science was never merely secular *en se*; it had to be *made* secular. Secularism proper contributed significantly to the secularizing of science in Britain.

I begin here by tracing a tradition of scientific naturalism undertaken by the predecessors of Secularism, including Holyoake himself, and continue by outlining the principles of Secularism as sketched by Holyoake in several formats and across four decades, which also amounts to a brief word history of the associated terms.[18] I then summarize the word history of scientific naturalism, briefly characterizing it as a philosophical framework for science. By juxtaposing these two creeds and their word histories, I aim to show their rhetorical and philosophical family resemblance, as well as their close social connections. I then trace the contours of the social network that linked Secularism and the scientific naturalists, showing that the scientific naturalists preferred Holyoake's Secularism as opposed to the Bradlaugh brand. I conclude with some remarks regarding the implications of this argument for the historiography of science and the question of secularization.

Artisan scientific naturalism

A form of scientific naturalism was promoted in artisan freethought circles from the early 1840s. Lamarckian transmutation theory played a major role in this freethought movement. In the 'hungry forties', evolutionary ideas were marshaled to counter a static, hierarchical, theocratic social order with a vision of a transformative, 'uprising' nature. An anti-theistic explanation for workings of nature was wielded to undercut the authority of the clerics and the basis of the state church. The malleability of the natural order spoke to the possibilities for changing social conditions.[19]

Beginning with the first number in November of 1841, the *Oracle of Reason*[20] included a serial article begun by Charles Southwell and continued by William Chilton entitled 'Theory of Regular Gradation', with woodcut illustrations of primitive man, fossils, and 'early' organisms. Serial publication lent itself well to the illustration of a theory of serial species change and development. As Secord notes, the first installment of 'Theory of Regular Gradation' began with an engraving of 'Fossil Man',[21] 'a racist fantasy lifted from the writings of the hack naturalist Pierre Boitard',[22] and representing 'man underdeveloped, as we are justified in supposing he was at that stage of his progress, when he was not exactly either monkey or man'.[23] The third installment began with a quote from William Lawrence, the materialist and the former professor of anatomy and physiology at the Royal College of Surgeons, whose expulsion had become a cause célèbre for the radical infidel, Richard Carlile.[24]

By the seventh installment of the article, with Charles Southwell imprisoned in Bristol jail for blasphemy in connection with the publication of his Christian-goading, anti-Semitic article 'The Jew Book' (see Chapter 3), Holyoake had taken over active editorship of the *Oracle*. William Chilton began authorship of 'Theory of Regular Gradation'. Chilton, whom Holyoake described as 'the only absolute atheist I have ever known',[25] immediately worked to establish first principles, arguing 'that the inherent properties of dull matter, as some bright portions of it have designated it, are good and sufficient to produce all the varied, complicated, and beautiful phenomena of the universe – however numerous the differences in other spheres may be in addition to those of our own ...'. The usual objections to materialism, Chilton argued, were based on an inadequate and impoverished conception of matter as 'dull' and inert. Instead, he saw matter as eternal and inherently possessing all of the properties necessary to produce its multifarious emanations, found throughout time and space:

> For believing matter to be infinitely extended, to be infinitely divisible, and capable of infinite combination or arrangement of the particles – we see no reason in flying to supernaturalism for an explanation of the ultimate causes which produce the results we witness.[26]

Chilton even dismissed the usual distinction between living and nonliving matter. Stones and crystals were 'alive'. They changed and evolved in the same sense as 'organic' matter.

For such materialists, matter was the sole creative force in the universe, capable of doing anything previously ascribed to God, including the production of new species. God was a phantasm invented to strip matter of its rightful throne. Following Lamarck, Chilton posited an inherent, *a priori*, teleological disposition in nature, a tendency toward complexity and progress, and proffered the Lamarckian notion of adaptation to changed conditions by species from the remotest ancestor to the present:

> it adapted itself to alterations in the surrounding circumstances which were continually taking place; and, in process of time, resulted in a form so distinct from the first, as, without the intermediate modifications, to warrant the supposition that it never could have been produced from, or had any connexion with it.[27]

As Secord notes, under Chilton's pen, in addition to general principles, 'Theory of Regular Gradation' included 'recondite details' from the works of 'Cuvier, Robert Grant and other authorities'.[28] The series began to follow Chambers's *Information for the People,* adducing some of the same source material that would be used in *Vestiges of the Natural History of Creation* (1844), but appropriating it for avowedly materialist, atheistic ends. For example, Chilton adopted the nebular hypothesis as found in *Information for the People*,[29] nevertheless ridiculing the admission by 'Messrs. Chambers' that they found the hypothesis 'new and startling'. Representing an evolutionary understanding of the development of solar systems, the nebular hypothesis served Chilton's evolutionary agenda. Such ideas, Chilton claimed, had been propounded 'years before, by the despised, insulted, and persecuted Infidel', well before they had been safely accepted and 'given to the world by *respectable* men'.[30]

The species transformism of the *Oracle* preceded the appearance of evolutionary thinking in Chambers's *Vestiges* by a few years.[31] In fact, by the time *Vestiges* had been published, Chilton had already mined many of the same sources that Chambers used for his evolutionary cosmogony. In the process, he virulently criticized 'the cowardice and dishonesty' of scientific men, and science publishers like Chambers, who failed to openly avow the atheistic implications of recent findings in the physical sciences. Rather than removing error from the public mind, they compounded error by the mixing of scientific fact with religious speculation.[32]

By 1844 and the publication of *Vestiges*, the *Oracle* had been superseded by the *Movement* and the 'Theory of Regular Gradation' had been discontinued. Before ending the series, Chilton apologized to readers of the *Oracle* for his apparent failure to engage his readers with the material. He admitted that he might have made the series unnecessarily dry and difficult. 'This course in other hands might have been fraught with beneficial results, but in my case I fear it has failed', he wrote in the thirty-eighth installment.[33] Chilton recognized *Vestiges* as a 'successful' version of his efforts when the treatise appeared only a year later.

Chilton's response to *Vestiges* corroborates Secord's claim that Chambers had 'domesticated' evolutionary theory by bringing it into the middle-class Victorian home.[34] While claiming that the work included 'nothing new', Chilton suggested that the treatise nevertheless presented evolutionary ideas that were 'new to the world' and thus had the potential to 'startle many a pedant from his slumbers'.[35] However, the freethinker had to draw out such potential. Chilton saw the publication as an opportunity to re-interpret evolutionary theory so that its radical implications could be made clearer, likewise undermining its domesticating effect. Two years after its publication, he continued to write about *Vestiges* in the *Reasoner,* criticizing both its theism as well the accusations of its critics, who insisted that its author was a materialist: 'The author of the "Vestiges" is no materialist. He looks through matter up to matter's god; he is, in fact, "a pure Theist."'[36] Holyoake also seized on the opportunity to use *Vestiges* as a vehicle for extending the reach of freethought. In 1845, he devoted a Sunday lecture to 'the origin of man as set forth in that extraordinary work just published, entitled Vestiges of the Natural History of Creation', and, as a letter from Chilton attests, he even planned to write a book on the topic.[37] While Holyoake never completed his digest of *Vestiges,* his consideration of the project speaks volumes of his desire to enter into other circles for the promulgation of freethinking ideas, which he would do from the late 1840s.

The principles and word history of secularism

Within two decades of its founding by George Jacob Holyoake in 1851–1852, although Holyoake was widely recognized as Secularism's founder and first leader (see Chapter 3), Secularism had come to be identified with Charles Bradlaugh and the National Secular Society (NSS), of which Bradlaugh was the first president. By the late 1860s, Holyoake had ceded, somewhat unwittingly, his former centrality in the movement. Further, he no longer maintained exclusive control of the term that he had coined to represent it. Holyoake's inability to hold sway over his neologism may be seen as parallel to Huxley's later difficulty with 'agnosticism', which Huxley had coined in 1869 to represent his own creed in the context of the Metaphysical Society.[38] Secularism, both the movement and the word, had slipped from Holyoake's grasp for several reasons. First, as we saw in Chapter 3, Holyoake alienated staunch freethinking atheists, who essentially refused his construal of Secularism, while they nevertheless operated under the rubric and remained important advocates for the movement. Confidence in Holyoake's leadership was undermined as his disputed business practices, aversion to centralized organization, and comparably measured rhetorical approach were criticized and challenged.[39] The founding of the secularist *National Reformer* in 1860, with Bradlaugh as co-editor, along with the establishment of the NSS in 1866 with Bradlaugh as president, did much to officially reduce Holyoake's prominence within Secularism. Further, the

Knowlton affair of 1877 calcified the rift between the Holyoake and Bradlaugh camps, evoking the censure of the latter by the former.[40] Yet this disapprobation was a consequence of the significant media attention paid to Bradlaugh and Annie Besant on the occasion of their trial for obscenity, which further associated Secularism with Bradlaugh. Bradlaugh's election to the House of Commons for Northampton in 1880 and his eventual seating in 1888 augmented his renown.[41] After the critical early years, Holyoake intervened on the behalf of Secularism on many occasions, for example to write the *Principles of Secularism Briefly Explained* in 1859, to pen *The Principles of Secularism* in 1870, to debate Bradlaugh in March of 1870, and with Charles Watts (Sr.), G. W. Foote, and others, to (unsuccessfully) challenge the presidency of the NSS in the wake of the Knowlton affair.[42] Despite these efforts, Secularism was often regarded in the terms provided by the older infidelity as adopted by Bradlaugh. Yet, as I will show, it was to Holyoake and his version of Secularism that the scientific naturalists looked for a respectable and useful example of freethought as they named, developed, and promoted their cosmology.

Parallel to Huxley's bid to regain command over the usage of 'agnosticism' with the three essays he wrote for *The Nineteenth Century* in 1889,[43] late in the century, Holyoake sought to reassert his priority where Secularism was concerned – to solidify his legacy as its founder, and, yet again, to insist upon its original principles. In 1896, in *English Secularism, A Confession of Belief*, he left a retrospective index of ten documents that he regarded as foundational for Secularism's inception and establishment.[44] Other than the first two articles, the Preface to the *Movement* and the lectures to the Manchester Order of Odd-fellows, the documents had been published in the *Reasoner*. Holyoake clearly demonstrated that his *Reasoner* had been at the center of the movement. He reminded readers that he wrote all of the foundational texts, other than those that were addressed to him: 'These citations from my own writings are sufficient to show the origin and nature of Secularism'.[45] While an exclusive textual focus is by no means sufficient for understanding the cultural meaning and significance of Secularism, these texts nevertheless testify to the essential character of the Secularist creed as Holyoake saw it. Further, such a reading represents an exercise in 'word history' or 'historical semantics'. As Dawson and Lightman point out, drawing on Thomas Dixon's *The Invention of Altruism* (2008), 'the relation between words and concepts is never simply neutral, and the changing fortunes of a term have significant implications for the construction and communication of the ideas it might entail'.[46] In the case of Secularism, the fate of the word involved its appropriation by others in the freethought movement and especially the larger Secular camp headed by Bradlaugh. This appropriation has had significant implications concerning the meaning and understanding of Secularism proper, and the meaning and significance of modern secularism in general. It has led to confusion such that modern secularism is understood primarily as the absence or negation of religion and belief.

The first principle of Holyoake's Secularism was materialism, as enunciated in the *Movement*: 'Materialism will be advanced as the only sound basis of rational thought and practice', which 'restricts itself the known, to the present, and ... to realise the life that is'.[47] The remaining points were made in the *Reasoner*, and included some of the first usages of the words 'Secular' and 'Secularism' as denoting and describing a new system of knowledge and morality. The twelfth volume of the *Reasoner* opened with an article entitled, 'Truths to Teach', which undertook to 'indicate some of the objects which this journal endeavors to explain and enforce'. The first two points had been made in the *Oracle* and the *Movement*, and in earlier volumes of the *Reasoner*:

1. To teach that Churches, in affirming the existence of a Being independent of Nature, affirm what they do not know themselves – that they who say they have discovered Deity assume to have found what he has evidently chosen to conceal from men in this life by endowing them with finite powers ... – that whoever bids us depend upon the fruition of a future life may betray us from the use of this world.
2. To teach men to limit, therefore as a matter of truth and certainty, their affirmations to what they know – to restrict, as a matter of self-defence, their expectations to that which their experience warrants.[48]

In this article, later recognized as foundational to the incipient Secularism, one of the *Reasoner's* stated aims was to set limits on knowledge claims. Such limits would involve the restriction of knowledge to 'that which experience warrants'. Theology was deemed a 'science of conjecture' in affirming what can only be believed without knowledge, given the 'finite powers' of the human faculties. With these principles, Holyoake sought to remove freethought from the field of conjecture, and to confine it, as stated in the second point, to matters of 'certainty', or what could be known given the limited faculties. Under this principle, science was deemed the sole 'Providence of Man', which could be relied upon as an insurance against 'false dependencies'.[49]

With this announcement of aims, the *Reasoner* did not make the denial of deity necessary for the would-be Secularist. Knowledge for the benefit of humanity was separated from conjecture, which had not proven its benefits in the realm of experience. The *Reasoner did* warn against the affirmation of deity and a future life, given that reliance on them might 'betray us from the use of this world' to the detriment of 'progress' and amelioration. However, it warned only that such conjecture should be left behind for the purposes of pursuing knowledge and improving material conditions. Likewise, belief was not a disqualification for the pursuit of knowledge or progress, only a possible obstacle. One's belief in the supernatural was a matter of speculation or opinion to which one was entitled, unless such belief precluded positive knowledge or action. This rhetorical and philosophical turn represented the cleanest break hitherto from the previous dogmatism of freethinking atheism, while also marking the nascent Secularism as a precursor of agnosticism and scientific naturalism.[50] While Holyoake was inconsistent on this point and included atheism as the 'negative aspect' of Secularism as late as 1854, as I have shown in Chapter 3, he reiterated the distinction between Secularism and freethinking atheism often. For example, in March of 1858, he argued that:

> [t]o make Atheism the Shibboleth of the Secular party would be to make Secularism an atheistic sectarianism as narrow and exclusive as any Christian Sectarianism. The principles of Secularism are distinct both from Atheism and Theism, and there can be no honest, useful, wide, and liberal party without keeping this point well understood.[51]

He later suggested that Secularism considered both theism and atheism as 'belonging to the debatable ground of speculation' with their 'theories of the origin of nature'. Secularism 'neither asks nor gives any opinion upon them, confining itself to the entirely independent field of study – the order of the universe'. Holyoake could note in hindsight that similarly, 'Huxley's term agnosticism implies a different thing [than atheism] – unknowingness without denial', but 'unknowingness without denial' was fundamental to Secularism from its inception.[52]

With the third object of 'Truths to Teach' – 'to teach men to see that the sum of all knowledge and duty is secular – that it pertains to this world alone'[53] – Holyoake could rightly claim to have been an innovator, if not a neologist; 'this was the first time the word "Secular" was applied as a general test of principles of conduct apart from spiritual considerations', Holyoake claimed.[54] The Secular principle was in effect an ontological demarcation stratagem, dividing the metaphysical, spiritual or eternal from 'this life' – the material, the worldly, or the temporal: 'Secularity draws the line of demarcation between the things of time and the things of eternity'.[55] The 'Secular' for Holyoake designated the only domain where knowledge could be gained and effective action taken.[56] Like Karl Popper's later demarcation of science from pseudoscience and metaphysics in the *Logic of Scientific Discovery* (1959),[57] Secularism deemed that whatever could not be 'tested by the experience of this life' should simply be of no concern to the scientist, moralist, or politician. The 'Secularist' was one who restricted efforts to 'that province of human duty which belongs to this life'.[58] According to Holyoake, this was the first time the word 'Secularist' was used to denote an adherent to a 'new way of thinking' – to represent one who avowed Secular principles.[59] In fact, as I point out in Chapter 3, W. H. Ashurst, writing to the *Reasoner* under the pseudonym 'Edward Search', first suggested the words 'Secular' and 'Secularist' to describe the new branch of freethought that Holyoake was developing, and one who aligned with it. In the same article, Holyoake coined the term 'Secularism' to describe 'the work we have always had in hand'.[60]

Secularism was advanced not only as an epistemology but also as a morality and politics. With his fourth aim, Holyoake argued for the 'independent origin' of morality. Rather than being based on religious doctrine, the source of morality was nature – 'the real nature' of human beings – and its warrants were to be found in the consequences of actions, 'natural sanctions of the most effective kind'.[61] Never a strict Benthamite, and harking back to the social environmentalism of Godwin and Owen, Holyoake based morality primarily on the purported goodness of human nature itself, and only secondarily, in conjunction with practical results. Without a basis of natural goodness, a Secular system would be unable to warrant motives for right actions. '*Its theory of Morals* – That there exist guarantees of Morality in Human Nature, in Utility, and Intelligence'.[62] Intelligence, an aspect of human nature developed by knowledge, was required in order to discriminate between good and deleterious effects. The results were evaluated by intelligence according to utilitarian ethics, which in turn resulted in moral knowledge that influenced future actions. Politics was simply morality writ large. Thus, a moral and political science was advanced, comprised of a guiding principle and a scientific method.

In its claims for a political science based on human nature, Secularism was similar to the Positivism of Auguste Comte. However, Holyoake never suggested, as did Comte, that once discovering the social laws, human beings must subject themselves to those laws in an act of acquiescence, which has been seen as Positivism's conservative character. For Comte, the laws for conduct were not necessarily in human nature alone, but in a 'social physics' based on human nature. Comte avowedly aimed at establishing a 'social physics' in order to avert social and political chaos by positing a social lawfulness consistent with physical regularity.[63]

The fifth point urged the trust of nothing but 'Reason' for the establishment of all knowledge. The concept of reason was, as usual, a very slippery one. Its meaning could really only be completely understood by reference to what it excluded – in all cases, religious and other metaphysical speculation. It was not primarily distinguished from imagination as in Romanticism, but rather from the unsubstantiated belief of theology. Reason was figured as the logical treatment of experience, relying on 'nothing which does not come within the range of phenomenon, or common consciousness, or assumes the form of a law'.[64] The point was to derive knowledge by means of the intellectual processing of empirical data as opposed to accepting *a priori* convictions.

Free inquiry and discussion comprised the sixth aim. Only those statements withstanding the test of 'universal free, fair and open discussion ... the highest test of vital truth ... can be trusted',[65] Holyoake argued. '[O]nly that theory which is submitted to that ordeal is to be regarded, as only that which endures it can be trusted'.[66] In the requirement that all propositions stand the test of criticism and 'testing', the sixth object resembles Popper's criterion for science – the subjection of statements to possible disqualification or falsification in an agonistic field of testing and discourse.

These principles represented the 'positive aspect' of Secularism. At least until 1854 and possibly later, Holyoake wavered slightly on the dividing line between Secularism and earlier freethought; Secularism's 'negative side', which was to 'protest against specific speculative error' (theism), was occasionally revived. The two sides sometimes remained together under Secularism as a '*double* protest'.[67] However, the tendency was to jettison the protest and to emphasize Secularism as a new kind or stage of freethought – that is, to assert Secularism's limitation to the field of positive knowledge and to posit a substantive morality, as opposed to or exclusive of the negation of deity and theology.

The creed and word history of 'scientific naturalism'

Since Frank M. Turner's groundbreaking work, *Between Science and Religion* (1974), 'scientific naturalism' has been understood to be central to the historiography of Victorian science and the development of modern science. However, Turner argued in a subsequent essay that scientific naturalism should not be understood strictly in terms of its epistemological challenge to 'theistic science'.[68] Importantly, he suggested, it served as a means for establishing cultural authority, and as the basis for the 'professionalization' of the sciences, and thus the remaking of the intellectual landscape of Britain. In *Contesting Cultural Authority* (1993), Turner extended this approach, treating the conflict between the scientific naturalists and theists across several domains and in connection with numerous issues.[69] Whether or not one accepts Turner's claims for professionalization, and there have been critics,[70] there is little disagreement that the contest in the second half of the nineteenth century in Britain was between an Anglican clergy who had dominated the intellectual terrain occupied by 'science' and who depended on a theistic-scientific worldview, and an emerging elite from outside of their ranks, who worked to establish new criteria for conducting science and who attempted to make the new criteria the *sine qua non* of scientific practice and theory. Scientific naturalism was a retrospective shibboleth used by Huxley to refer to the epistemological, cultural, and professional identity values of this latter group.

Huxley chose the term scientific naturalism to replace his earlier coinage, 'agnosticism', most likely because he had lost control of the latter to Herbert Spencer and others. As Lightman has argued, Huxley had particularly lamented its slippage into usage by the son of Charles Watts (Sr.), Charles Albert Watts. The latter deployed the term (and the contents of a private letter from Huxley) for the 1884 launch of his agnostic publication, the *Agnostic Annual*, in order to sanitize freethought of its working-class, atheistic taint in an effort to gain respectability for a newly energized and broadened secular movement.[71] Scientific naturalism thus became Huxley's phrase of choice to represent the creed, and 'scientific naturalist', a term with an even longer history, was taken to represent one who held it.

Huxley first publicly used the term 'scientific Naturalism' in 1892 in the prologue to his *Essays Upon Some Controverted Questions* to refer to the scientific worldview that he and his fellow supporters of Darwin had espoused for several decades. In so doing, 'Huxley [apparently] introduced another new term, akin to his earlier coinage *agnosticism*, into the lexicon of nineteenth-century science'. In fact, as Gowan Dawson and Bernard Lightman have discovered, a professor of Greek at Union College in Schenectady named Tayler Lewis had used the expression as early as the late 1840s. Writing in the American evangelical press, Lewis disparagingly referred to a 'merely scientific naturalism', a scientific worldview that discounted the continuous superintendence of a 'Supernatural Power' over the physical universe. The phrase subsequently entered into the British lexicon in the 1860s, where it was also deployed pejoratively in the religious press to refer to a scientific worldview unmoored from theology.[72]

The first sympathetic use of scientific naturalism came, perhaps unsurprisingly, in the pages of the *Secular Review*,[73] a magazine launched by Holyoake in 1876, one in a long line of freethought publications with which he was associated. Holyoake soon handed over the proprietorship of the publication to Charles Watts (Sr.), and by June 1877, with Charles Watts (Sr.) and G. W. Foote as editors, the periodical had become the *Secular Review and Secularist*, the organ of the new breakaway British Secular Union (BSU).[74] In the aftermath of the Knowlton affair (see Chapter 3), Holyoake, Watts, and Foote, among others, lent the BSU and the *Secular Review* support in a campaign to establish an organization of Secularism apart from the NSS.

In February 1878, in response to a correspondent to the *Secular Review* writing under the pseudonym 'Naturalist', who had argued that 'Spiritualism is a phase of Naturalism',[75] 'Draco' referred to 'Scientific Naturalism' as the worldview held by those who had eliminated spiritualism and supernaturalism from their science:

> In reading through the 'Open Column' of your well-conducted journal, my attention was drawn to a letter, entitled 'The Plea of a Convert', by 'Naturalist' ... I do not think that it is possible to point at one gentleman who pertains to eminence in physical science who does acknowledge the truth of this modern phenomena [spiritualism] farther than from natural principles. If 'Naturalist' holds on to his plea in the dark, he must drop the cognomen 'Naturalist', for a true Spiritualist grasps the supermundane principle ... I presume to say for myself – being a reader of the current literature for and against this new malady – that men of science have laid the subject well open, and that they find that the immaterial element is all nonsense. Not but that these men have something in hand to grapple with; fact after fact presents itself in the phenomena; but it all ends in *natural* science. If I mistake not, I think 'Naturalist' is trying to kick over the traces of Spiritualism, and wants to make a couplet of the two – Spiritualism and Naturalism. But they cannot clash together under the idea of *Scientific Naturalism*.[76]

In November 1876, John Tyndall had written to Holyoake to take out a three-year subscription to the *Secular Review*.[77] Perhaps Tyndall, Huxley's coadjutor in the fight against theistic science, had informed his friend of this favorable terminological usage. In any case, Huxley somehow discovered the phrase, finding it appropriate and useful for his purposes. However, he would first have to strip it of the supposed 'grubby taint of Watts's secular penny press', while also disassociating it from 'its awkward origins as a pejorative label employed by evangelicals on both sides of the Atlantic'.[78] He did so upon first mention in the 1892 prologue:

> It is important to note that the principle of the *scientific Naturalism* of the latter half of the nineteenth century, in which the intellectual movement of the Renascence has culminated, and which was first clearly formulated by Descartes, leads not to the denial of the existence of any Supernature; but simply to the denial of the validity of the evidence adduced in favor of this, or of that, extant form of Supernaturalism.
>
> Looking at the matter from the most rigidly scientific point of view, the assumption that, amidst the myriads of worlds scattered through endless space, there can be no intelligence, as much greater than man's as his is greater than a blackbeetle's; no being endowed with powers of influencing the course of nature as much greater than his, as his is greater than a snails, seems to me not merely baseless, but impertinent.[79]

Scientific naturalism simply made no claims about the existence of the supernatural; the 'man of science' 'taking refuge in that "agnostic" confession, which appears to be the only position for people who object to say that they know what they are quite aware they do not know'.[80] That is, like agnosticism before it, and Secularism before that, scientific naturalism was a *methodological* rather than an ontological or metaphysical naturalism.[81] It did not require the denial of the supernatural, but merely excluded it from scientific explanation. Theoretically, the scientific naturalist could be both an ontological theist and a methodological naturalist; just as the Secularist, under Holyoake's formulation, could be both a theist and a methodological materialist. (See Chapter 3.)

Careful to distinguish between a noble lineage for scientific naturalism and a crass, (French) Enlightenment materialism, Huxley continued by insisting on the creed's constructive aspect:

> I have hitherto dwelt upon scientific Naturalism chiefly in its critical and destructive aspect. But the present incarnation of the spirit of the Renascence differs from its predecessor in the eighteenth century, in that it builds up, as well as pulls down.[82]

Huxley then erected the likely pillars of this secular creed, exhorting 'that all future philosophical and theological speculations will have to accommodate themselves to some such common body of established truths as the following'. These propositions included the fundamental facts of organic life, from the long-term existence of plants and animals on the planet, to the simplicity with which all organisms commence life and then differentiate into complex forms, to the unconsciousness of plants, to the likely beginnings of consciousness, to the animal ancestry of humanity, to social organization as the basis for morality, to the highest form of human society being that in which consideration of the whole of society dominates. Much like Holyoake's Secularism, that is, scientific naturalism would consist of a set of positive propositions, rather than mere negation. It is telling that the fundamental truths discovered by this scientific naturalism numbered twelve, and that Huxley ended the prologue with a plea to retain the Bible as a source for popular education.

Secularism and scientific naturalism: a social network

Gowan Dawson notes in his book on Darwin and respectability that while 'Darwin's deliberate and often rather haughty eschewal' of the Bradlaugh wing of Secularism is well known, 'the simultaneous endeavors of some of his principal supporters, including Huxley and John Tyndall, to forge closer connections with those free-thinkers and radicals' in the Holyoake camp of Secularism 'are less well known'. The 'complex negotiations' with such freethinkers 'were crucial to the endeavor to establish Darwinism', Dawson continues.[83] Following up on this rather redolent hint, in this section, I trace some of these complex negotiations and the forging of connections between the scientific naturalists and the Holyoake camp of Secularism.

As we have seen, the relationship between Holyoake's Secularism and scientific naturalism consisted of a philosophical family resemblance and a link in the Secular press. But it also entailed a longstanding communications network and mutual support system that began at mid-century. In the early 1850s, in connection with the new Secularist circle surrounding the *Reasoner* and the *Leader*, Holyoake had made contact with Spencer, and probably Huxley. Huxley, Spencer, and Tyndall surely knew of Holyoake's brand of Secularism then in circulation, and it is possible that Spencer and Huxley were involved in the 'Confidential Combination' that discussed and helped to develop it. By the 1860s, as correspondence and other evidence shows, Holyoake had earlier secured the confidence of these leading lights as they forged ahead in establishing the relative dominance of scientific naturalism. Holyoake had proved himself a trustworthy figure, 'the former firebrand'[84] whose opinions and behavior were becoming congruent with middle-class morality (or elite culture),[85] a culture that was undergoing significant change, due in no small part to the impact of scientific naturalism itself. Holyoake's Secularism could not be mistaken for the old infidelity, and Holyoake conveniently used Bradlaugh's atheism and neo-Malthusianism as a foil to differentiate his strand. The scientific naturalists offered their support to Holyoake and his camp, in return for Holyoake's fidelity to a respectable brand of freethought and for providing a safe bridge back to working-class unbelief for the promotion of scientific naturalism and evolutionary science.

From as early 1860, until the end of the century, Holyoake regularly corresponded with Huxley, Spencer, and Tyndall. The letters, which number in the several dozens, covered numerous issues, including polemics against religious interlocutors,[86] the mutual promotion of literature,[87] the naturalists' financial and written support for Secularism and Secularists,[88] and health,[89] amongst other topics. Likewise, while he is correct in assessing the importance of the Holyoake camp to scientific naturalism, Dawson is mistaken when he suggests that the relationship was based exclusively on agreement over birth control and sexual policy. According to his interpretation, the fallout occasioned by the republication and legal defense of Knowlton's *The Fruits of Philosophy* in 1877 by Charles Bradlaugh and Annie Besant became the primary reason for the division between the Holyoake and Bradlaugh camps. Birth control and sexual policy, Dawson argues, 'were by far the most divisive issue[s] within the British freethought movement in the nineteenth century'. Dawson suggests that the distinction between what Michael Mason refers to as the 'anti-sensual progressive' (Holyoake) and the 'pro-sensual' (Bradlaugh) Secularist camps was the sole basis for the differential esteem accorded the two camps by the Darwinian circle. Darwin, Huxley, Tyndall, and others, who 'vehemently opposed any attempts by radicals to appropriate evolutionary theory to justify their support for contraception', deplored Bradlaugh's 'neo-Malthusian' position.[90] They found Holyoake acceptable due to his compatible sexual policies.

While Dawson is right to suggest that the Darwinian circle strongly preferred Holyoake's position regarding Bradlaugh's and Besant's republication and legal defense of Knowlton's *Fruits of Philosophy*, his interpretation fails to account for the earlier relationships between Holyoake and the scientific naturalists, while it misses the fundamental division within Secularism. First, as I have suggested in Chapter 3, the primary split dated to the early 1850s and went to the definition of Secularism itself. In 1957, G. H. Taylor, a chronicler of Secularism, and obvious follower in the Bradlaugh lineage, put it as follows:

> Is the theoretical attack [on Christianity] really necessary or advisable? That was the problem which did more than any other single factor to split the ranks. Roughly speaking, Holyoake said No, Bradlaugh, Yes ... Moreover, no thoroughgoing secularist can subscribe to Holyoake's admiration of Comte's Positivism, which has been called Roman Catholicism minus Christianity. It is only fair to add that a case can be made for secularists getting on with the job without unduly antagonising their potential supporters with such shocking heresies as Atheism and the denial of survival after death, not to mention the exposure of Bible absurdities.[91]

As we can see, the initial rift was not over sexual policy but rather principle and strategy regarding religion and belief. In fact, years before the Knowlton affair, Holyoake had effectively denied that Bradlaugh was a Secularist at all.[92] Bradlaugh and company insisted on atheism as an essential conviction for the Secularist and bitterly reproached Holyoake and his followers for their conciliation with theists. Indeed, Bradlaugh's rise had much to do with the trenchant atheistic and anticlerical rhetoric he and others conducted in the *National Reformer*[93] and the *Investigator* before it.

This distinction drew the scientific naturalists to Holyoake's side, as Holyoake's Secularism served the scientific naturalists well. For instance, as Huxley struggled to dissociate himself from charges of materialism and atheism throughout his polemical career,[94] Holyoake provided ready assistance. In April 1873, four years before the Knowlton affair, Huxley wrote to Holyoake:

> I am too lazy to defend myself against injustice although I am all the more obliged to men who are generous enough to take the tumble for me – so I offer you my best thanks for your successes [in arguing] against [Moncure] Conway's association of me with Bradlaugh & Co. – for whom & all their ways and works I have a peculiar abhorrence.[95]

Such 'peculiar abhorrence' was expressed for 'the coarse atheistic philosophy of Bradlaugh and his secularists [which] had always repelled Huxley and many of his scientific naturalist colleagues'.[96]

Nevertheless, the secession of Holyoake, Charles Watts Sr., G. W. Foote, and other freethought radicals from the NSS, and their founding of the BSU in August 1877 following the Knowlton affair, certainly did much to cement relations between the Holyoake Secularist wing and the Darwinian naturalists. In July 1877, when the controversy was raging, Tyndall wrote to Holyoake thanking him for a clipping from the *Birmingham Weekly*, in which Holyoake likely denied having formerly published *Fruits of Philosophy*,[97] adding the remark: 'I do not agree with you in all political things, but I have always recognized your straightforwardness and truth'.[98] Further, in his Presidential Address to the Birmingham Midland Institute in October 1877, Tyndall extolled Holyoake as an exemplar of secular morality:

> To many of you the name of George Jacob Holyoake is doubtless familiar, and you are probably aware that at no man in England has the term 'atheist' been more frequently pelted. There are, moreover, really few who have more completely liberated themselves from theologic notions. Among working-class politicians Mr. Holyoake is a leader. Does he exhort his followers to 'Eat and drink, for to-morrow we die?' Not so.[99]

In the *Secular Review*, where the writing and speeches of the scientific naturalists were regularly reported, the Secularists of the Holyoake camp celebrated this acknowledgement of Holyoake by Tyndall:

> I, in common with many others, have been gratified that Mr. G. J. Holyoake's style of advocacy has received such marked approval as that accorded it by Professor Tyndall at Birmingham. To receive such a notice, on such an occasion, from one so eminent is indeed an honour that goes far to repay years of toil and study, and compensates for much misrepresentation and abuse.
>
> Mr. Holyoake has lived long enough to formulate into a system the views he espoused so early, and has advocated so long; has seen them attain a considerable share of popularity, and now has the satisfaction of having one phase of his teachings publicly approved by one of the foremost men of his time. Long may he be spared to receive many such honours, to inspire us with his fidelity, to direct us by his counsels, and to benefit us by his example.[100]

From the 1860s through the 1870s, as a correspondent for British and American radical presses, Holyoake reported on meetings of the

British Association for the Advancement of Science, just as the scientific naturalists emerged as dominant players.[101] In 1870, when Huxley became the President of the BAAS and presided over the proceedings at Newcastle-upon-Tyne, Holyoake, who was covering the meeting, wrote to Huxley to complain about the treatment of the press. 'If science, as Prince Albert said is to be popular, you know the press is one of its agents. No association treats the press more coldly than the Brit. Assn. ...'. He lamented the 'Reporter Admission' (£15) and the BAAS's refusal to give him a copy of Huxley's Presidential Address: 'What can be the cost of a copy – even if the applicant known to be of the press misused it – compared with the service which as a rule they render?' Holyoake ended by promising to 'celebrate your [Huxley's] reign to the ends of the earth', if only he would 'mitigate this ignominious parsimony'.[102]

It is tempting to consider the treatment of Holyoake by the BAAS as premised upon his reputation as a former radical atheist and in terms of a bias against the press outlets for which he was reporting. Yet, at the BAAS Annual Meeting in 1867 at Dundee, Sir John Lubbock had praised Holyoake, who was present as a reporter. During a session at the meeting over which he presided, Lubbock thanked Holyoake for the services the latter had rendered freethought: 'The baronet declared, that but not for the labors of Mr. Holyoake, it might not have been possible for them, the *savans*, to speak as freely as they do in these days'.[103] Returning the favor, Holyoake, reporting on the meeting for the *New York Tribune*, celebrated Tyndall's 'materialism' and noted the consternation of the new Chair of the BAAS, the Duke of Buccleuh, during Tyndall's address.[104]

The relationships between Holyoake and the members of the Darwinian circle could also be quite personal and involved support during illness. In April 1875, Spencer asked for a copy of Holyoake's *History of Cooperation*, offered his condolences to Holyoake for the latter's (temporary) blindness, and promised to contribute to the fund established to support him during his indisposition.[105] In the same month, Tyndall also responded, writing to Evans Bell, who had established the fund:

> Permit me to say that I have received with genuine sorrow the intelligence it conveyed of Mr Holyoake's failing health. And allow me also to thank you for giving me the opportunity of showing, even in the smallest way here open to me, my appreciation of the character of one upon whose life is stamped, with singular distinctness, the image and superscription of 'an honest man'.[106]

In November 1875, Huxley wrote to Holyoake, wishing him 'with all my heart a speedy return to the visible world – which is on the whole a pleasant spectacle'. He also contributed to the fund for the man, 'who has so long & so faithfully served the cause of Free thought'.[107]

As he often liked to remind readers and audiences, Holyoake had earlier paid a price for freethought advocacy that the secularists, scientific naturalists, agnostics, and rationalists of the later part of the century would never have to pay.[108] Along with other freethinkers of lower-class origin, Holyoake had paved the way by fighting battles with the clergy, in the press and in lectures, long before Huxley took up this charge.[109] Yet, he continued to engage in the ongoing fray, defending Huxley, Tyndall, and others against their religious antagonists.[110] He served Huxley religiously when Huxley complained that controversy was 'hard upon a poor man who has retired to "make his sowl" as the Irish say, in the sea side hermitage'.[111]

In a eulogy to Tyndall, Holyoake claimed that Tyndall had paid him what was perhaps the highest compliment that the scientific naturalists could have rendered him publicly. Given Tyndall's earlier public and private statements regarding Holyoake, we have no reason to doubt Holyoake's veracity:

> I remember meeting Tyndall one day in Dundee, when the British Association for the advancement of science met there. The Duke of Buccleuh was President. Narrow-minded, of little knowledge, and possessing a larger share than was due to him of Scottish intolerance, the Duke had a bad time in the chair while Tyndall was addressing the saints and philosophers assembled. When the meeting was over I said to Tyndall, 'It's very well for you, you have come to Dundee late; the Duke's ancestors would, and I think he would, treat you like a witch, and try the persecution of fire upon you'. 'Ah! Holyoake', he replied, 'it's very well you went before us. We do but gather where you have sown'.[112]

In Holyoake's account, Tyndall pointed to Holyoake's efforts in the freethought movement – his battles for free expression, his trial and imprisonment for blasphemy, and finally, his partial victories against religious prejudice – for making possible the open expression of naturalistic views, and perhaps for having helped to formulate them. As we have seen, similar remarks were made publicly by John Lubbock at a meeting of the BAAS in 1867.

By the 1870s, gentlemen freethinkers no longer felt compelled to meet in secret clubs like Hunt's Confidential Combination. Instead, the liberal theologian Moncure Conway founded an 'Association of Liberal Thinkers' in June 1878, boasting of 'the first effort ever made to unite persons interested in the religious sentiment and the moral welfare of mankind on a plan absolutely free from considerations of dogma, race, names, or shibboleths'.[113] Holyoake might have found such a declaration galling, given his decades-long attempt to do the same with Secularism. The club's purpose was to bring together for candid discussion men of various political, theological and philosophical positions. Huxley was elected President, while Tyndall, Clifford, Kalisch and Holyoake were nominated Vice Presidents.[114] Darwin was invited to join, and although he sent a donation, 'ally himself publicly with organized freethought he would not'.[115] The organization was short-lived, lasting a mere six months before collapsing for lack of a common mission. But Holyoake's inclusion and nomination shows the extent to which he had become acceptable in respectable circles. A similar club had met in 1873, apparently for the same purpose. It included 'Catholics, High Churchmen, Broad Churchmen, Dissenters, Come-Outers, Infidels, Positivists, Materialists, Spiritualists, and Atheists'. Members included 'Dr. [John Henry] Newman, Archbishop Manning, Dr. Pusey, Mr. Gladstone, Maurice, Huxley, Mill, Lewes, Bishop Wilberforce, and Mr. Holyoake'. The question for discussion in one meeting was, 'Is There a God?'.[116] In both societies, unbelief was now a question that could be vetted in polite company, and Holyoake was invited.

Holyoake's contribution to freethought and scientific naturalism would be incomplete without reference to the Rationalist Press Association (RPA), founded in 1899 by Charles Albert Watts, the son of Secularist Charles Watts. Despite serving as the Chairman of the Board of the RPA until his death in 1906, Holyoake's direct role in the organization was limited. But it begins with his association with the older Watts and Watts's split with Bradlaugh after the Knowlton trial.[117] The younger Watts converted his antipathy for Bradlaugh and his experience as a printer and businessman into a publishing venture, the RPA.[118] Leaving the editorship of the *Secular Review* (founded in 1876) to William Stewart Ross in 1883 and taking charge of Watts & Company, Charles Albert Watts launched the *Agnostic Annual* in 1884.[119] The annual was modeled on the symposium format introduced in the *Nineteenth Century* by James Knowles seven years earlier. Like Knowles, the younger Watts sought to enlist prestigious writers. Watts secured the contributions of Francis Newman, Leslie Stephen, Edward Clodd, Ernst Haeckel, and Thomas Huxley, among others.[120]

Bernard Lightman has treated this publishing venture and the popularizers who disseminated Darwinian ideas to the cloth-cap audience of 'new agnostics'. Bill Cooke has written a centenary history of the RPA, suggesting that Charles Albert Watts's business acumen and amicable personality enabled him to identify a new market niche that included both middle- and working-class readers and successfully to target them with a new class of mass freethought publications that included contributions from eminent, respectable writers.[121] This publishing trajectory parallels that of the atheist Darwinian popularizers derived from the Bradlaugh wing, in particular Edward A. Aveling, the socialist who branded his form of Darwinism for the members of the NSS through the Freethought Publishing Company.[122] But the RPA, as Royle points out, was 'the biggest breakthrough of all in freethought publishing', a remarkably successful printing, publishing, and propagandist venture that earned money and 'secured the pens of the leading figures of the day, including T.H. Huxley'.[123]

The RPA grew from the wing of Secularism that Holyoake had founded and tended. It stemmed from Holyoake's first publisher, Watts & Co. The RPA's audience included a new class of educated workers who benefited from National Education undertaken on a secular basis. The new liberalism had been born and Watts Jr. successfully identified and catered to the market. Watts eschewed not only the kind of bombast and negation expected from the NSS but also the internecine squabbles characteristic of Secularism's history.[124] Many of the RPA writers despised atheism more than they did theology, and, after the example of Francis Newman, valued religious sentiment as a part of human culture. RPA publications mostly avoided radical politics as well. As Royle notes, 'this was respectable freethought indeed, and the market was largely that toward which G. J. Holyoake had struggled, somewhat prematurely, in vain'.[125] According to Lightman, the RPA was one of the major means of delivering scientific naturalism and Darwinian evolution to mass audiences.

But what was the relationship between the RPA and the scientific naturalists, and how does this show a connection to Holyoake and his Secularism? Although Spencer, Clifford and Huxley had works reprinted by the RPA, Huxley's publication record with the RPA provides the most curious and illustrative example.[126] It begins with an egregious breach of publishing ethics by C. A. Watts[127] and continues with the publication of Huxley's later works with the approval of Huxley himself, and posthumously, with the approval of his son, Leonard, and his widow, Henrietta.[128]

In September 1883, Watts had written to Huxley to announce the forthcoming publication of his new *Agnostic Annual*, asking Huxley's 'advice and assistance because I am convinced that you are desirous of guiding and influencing the thought of the nation, to which you have already rendered such incalculable service'.[129] Misunderstanding or deliberately misrepresenting Huxley's intentions, Watts took writing that Huxley considered strictly private correspondence and published it in the first edition of the first volume of the *Agnostic Annual.* Huxley complained bitterly to Watts in a series of letters, while also expressing his outrage in other correspondence and in letters to the press. In late 1883, Huxley wrote to Tyndall about Watts's 'impudence' for 'printing this without asking leave or sending a proof, but paraded me as a "contributor"'.[130] Worse yet, after the first edition of Volume One quickly sold out, Watts also published the series of letters between himself and Huxley in the second edition of Volume One, in an attempt to vindicate himself and the RPA.[131] But in 1892, Huxley apparently entrusted Watts with his 'Possibilities and Impossibilities' for the 1892 volume of the annual,[132] and in 1902, Leonard Huxley successfully lobbied the Macmillan publishing house, which had rights in Huxley's *Lectures and Essays,* to grant the RPA permission to reprint the collection. By 1905, the RPA reprint had run to twenty editions and sold seven hundred and fifty thousand copies, making it one of the RPA's most successful publications.[133] Cooke argues that 'Leonard Huxley's willingness to ensure his father's work was reprinted by the RPA is further evidence that his father had relented of his early low opinion of Watts'.[134]

Adrian Desmond suggests that the pioneering young Watts exploited Huxley's coinage for his new *Agnostic Annual* in order 'to trade on agnosticism's respectability' and its promise of 'intellectual upward mobility', and that 'Huxley lost control [of the word "agnostic"] as the monthly Agnostic in 1885 preceded a spate of books capitalizing on the need for agnostic texts'.[135] Yet, as Cooke points out, this interpretation leaves unexplained Huxley's later, apparently voluntary involvement with the RPA, as well as that of his son and widow.[136] The apparent contradiction may be explained partly in terms of the RPA's differentiation from the Bradlaugh camp. After all, Huxley would never have agreed, under any conditions, to put his name on a Freethought Publishing Company text. Any prejudice that Huxley might have held with reference to the RPA would have been exacerbated by the initial publication debacle. Yet, Watts was later able to rescue his relationship with Huxley, to secure his work, and forge connections with his son and widow. Leonard Huxley even became an honorary associate of the RPA beginning in 1902.[137] Thus, it appears that Thomas Huxley's inclination was to work with the RPA.

In May of 1884, Huxley wrote to Holyoake on the Watts episode, referring specifically to Watts's use of his private correspondence:

> Many thanks for your note & enclosure. I was very wroth with Mr. Watts at the issue ... I wish every one had given as complete a refutation to the ... doctrine that freethinkers [can] 'make free' in their ways ... as per always [you] have my word indeed.[138]

Although the enclosure is missing from the record, Holyoake's habit had been to include clippings of articles in which he had supported the scientific naturalists in the press.[139] Likewise, it is apparent that Huxley was congratulating Holyoake for his 'refutation' of the 'doctrine' of making free with others' words. Watts's ability to secure Huxley's later work was due, in part, to the confidence that Holyoake's nominal involvement in the RPA provided. It appears that the network connecting Holyoake's Secularism and the scientific naturalists extended as far as the beginning of the twentieth century.

Conclusion: secularism, historiography, and 'secularization'

With mid-century Secularism, a cultural and intellectual work was done that contributed significantly to what James R. Moore has called the new 'creed' of scientific naturalism. By claiming to exclude questions of belief from those of positive knowledge, Secularism served as a precursor for advancing a naturalistic epistemology within science, thus addressing the 'science versus religion' controversy, however understood. Secularism paved the way for a partial détente between belief and unbelief that would be characteristic of agnosticism as a disposition of the later scientific naturalism. Secularism's contribution of an early form of agnosticism, I have suggested, did much to advance the worldview developed and promulgated by Huxley, Spencer, and Tyndall. The social network established between Holyoake and the scientific naturalists is further evidence of the compatibility of Secularism and scientific naturalism, and the work that the former had done for the latter, and vice versa.

What then are the implications of these findings for the historiography of science, for the relationship between science and the secular, and for the question of 'secularization' itself? In terms of historiography, non-elite, lower-class actors gain importance by this history. The focus on such non-elites has been a preoccupation of historians of science since Adrian Desmond's 1987 essay 'Artisan Resistance', and his major intervention with the *Politics of Evolution* (1989).[140] Here, non-elites gain importance for the *ethos* of science, rather than its actual practice.

Further, the connection between artisan Secularism and scientific naturalism sheds light on the class-character and origins of scientific naturalism itself. The scientific naturalists, it should be recalled, were young men during the 1840s. Both Tyndall and Huxley hailed from lower-middle-class backgrounds. Tyndall went to mechanics institutes regularly. Holyoake was read by a sophisticated working-class and lower-middle-class audience, a group to which Tyndall, Huxley, and even Spencer belonged in the 1840s, before they rose to prominence. Secularism and pre-Secularism doubtless would have been familiar to them, especially to Spencer as part of the Confidential Combination, and Huxley as he became part of the *Westminster* coterie at 142 Strand and the *Westminster* reported on Secularism in its pages.

More importantly, perhaps, this history has shown the danger of taking elite actors at their word. While occasionally recognizing Holyoake's role in tilling the soil for scientific naturalism – for easing legal restrictions and mitigating the moral opprobrium associated with freethought – the scientific naturalists rarely if ever paid tribute to Secularism's theoretical and philosophical contributions. It is important to recognize the reasons for this apparent neglect. Holyoake had advanced a form of 'agnosticism' well before Huxley coined the term in 1869, or appropriated 'scientific naturalism' from the Secular press in 1892. Yet, had he paid homage to a source associated with lower-class atheistic freethought and infidelity, Huxley would have tainted the new creed, and likewise undermined the very reasons for issuing a new terminology in the first place. Instead of crediting such a proximate source as Holyoake's Secularism, Huxley instead recalled a noble tradition dating to the Renaissance and extending to the Enlightenment, invoking the names of Descartes, Hume, and Kant. Certainly, such a 'grubby' name as Holyoake could not be mentioned in this connection. That does not mean that Secularism did not in fact contribute to the new creed. In an effort to achieve a most favourable self-representation, the rhetorical maneuverings of elite actors can elide important aspects of their own backgrounds, including cultural and intellectual debts. And historians of science can repeat these representations.

More importantly for my concerns here, our understanding of science's relationship to the secular is nuanced by this exploration. Even before considering Secularism as a source for scientific naturalism, it has been apparent in the historiography that the scientific naturalists did a great deal of cultural work to 'secularize' science and the broader social order. As numerous studies have made clear, this was a contingent, lengthy, and arduous campaign, and it was not a complete success. Now, given this prehistory of scientific naturalism in Secularism, we are made even more keenly aware of the contingent character of the secular in science, and that its emergence and success was anything but inevitable.

Whether or not the epistemology of scientific naturalism changed the way science was actually practiced is another matter. If, that is, with Matthew Stanley, we recognize that the theistic science practitioners and the scientific naturalists actually shared the same the operational assumptions (the uniformity of nature), and that epistemological or cosmological differences did not materially affect scientific practice,[141] we must recognize that scientific naturalism functioned primarily as an ideology for the promotion of its adherents and their objectives, rather than as a crucially distinctive feature of science. Given this, our conviction that science is not necessarily secular or secularizing is made even stronger.

Finally, this chapter sheds light on just what 'secularization' has meant in the context of science. Like its forebear in Secularism, scientific naturalism was a methodological rather than an ontological naturalism, an agnosticism that purportedly made room for metaphysical speculation, private belief, and even doctrinal Christianity. The supernatural was simply deemed irrelevant and the question of metaphysics was wholly inadmissible in the practice of science. As such, we might say that along with Secularism, scientific naturalism maps very well onto the notion of secularity advanced throughout; theoretically, belief and nonbelief/unbelief could be present in the same moment and even in the same person. As long as speculations and beliefs did not interfere with the work at hand, that is. For Secularism, the work involved, above all, efforts undertaken in 'this life' for the improvement of the observable social order based on a naturalistic epistemology and morality. For scientific naturalism, the work included the promotion of a particular vision for science and a particular (naturalistic) worldview for the broader social body.

5
The Three Newmans: A Triumvirate of Secularity

The widely divergent religious reconversions of the two Newman brothers – John Henry to Roman Catholicism and Francis William to a 'primitive Christianity' – have intrigued and troubled commentators and historians since the mid-nineteenth-century, when their religious crises represented a national concern. With the publication of Francis William Newman's *The Soul* (1849) and *Phases of Faith* (1850), at least one reviewer lamented the loss of two of the Church of England's most talented scholars and brothers – the one to 'superstition' and the other to 'unbelief'.[1] The fascination with the religious lives of 'the Newman brothers' continued well into the twentieth century and became the title of a book.[2] Even broader studies of the period included discussion of the two Newmans. For example, in *More Nineteenth-Century Studies: A Group of Honest Doubters* (1956), Basil Willey argued that the separate religious paths of Francis William Newman and John Henry Newman represented an important and telling fact about the nineteenth century in Britain, a fact that epitomized the religious ferment of the century itself:

> In the history of nineteenth century English thought there is no story more striking, or more full of moral significance, than that of the divergent courses of the brothers Newman. It is as if two rivers, taking their rise in the same dividing range, should yet be deflected by some minute original irregularity of level, so that one pours its waters into the Mediterranean, the other into the German Ocean ... The foundations of nineteenth century Protestantism were indeed insecure, and out of this insecurity there was bound to emerge, and did emerge, a drift on the one hand towards Rome and on the other towards unbelief.[3]

While Willey's characterization of Francis's spiritual destination as 'unbelief' missed the mark and his claim about Protestantism at large represented a rash over-generalization, the sketch nevertheless registered an important historical fact about nineteenth-century religiosity. The possibility of reaching such antithetical conclusions after beginning from the same evangelical and familial base remained a curiosity and a puzzle. I maintain that it is explicable in terms of the optative condition of secularity that I have discussed throughout.

Yet, while '[m]ost people were under the impression there were only two brothers, who had long figured in the public eye as types of the opposite courses of modern thought towards Romanism and Rationalism',[4] the two Newmans are only part of the complete Newman story. Excluded by this standard pairing has been the *third* Newman brother, Charles Robert Newman (1802–1884), who has been almost utterly neglected or has remained unfamiliar to historians.[5] Unlike his brothers, Charles Robert Newman was a relative unknown (and an unknown relative) even during his own lifetime. Although certainly making much less of an impression than the other two, Charles Robert is nevertheless necessary for completing the sketch of secularity that I am drawing here. Beginning also as an evangelical, Charles Robert represented a third possible destination from the common religious and familial base. His religious trek ended in committed atheism; this Newman thus represents the real antipode to John Henry: 'Yet the real type of antagonism to Rome was to be found in Charles Robert, who is dismissed by the Rev. Thomas Mozley [his brother-in-law] with the words: "There was also another brother, not without his share in the heritage of natural gifts"'.[6] For the most part, Charles was a *fait néant*, unable to display his natural gifts; after several aborted careers, he spent his last decades as a recluse and his manuscripts were destroyed upon his death.[7] Yet a record of his thinking does survive in the few writings that he sent to George Holyoake for the *Reasoner*, collected later in a short volume published posthumously in 1891.[8]

Although I will be referencing all three Newman brothers and rescuing at least one from near total oblivion, by far the central concern of this chapter is Francis William Newman, a figure who, of the three brothers, appears to have taken the *via media* between the ultra-orthodoxy of John Henry and the complete renunciation of belief of Charles Robert (although, as we shall see, Francis's was a far left middle road that was extremely controversial and attracted its share of vehement criticism). While together the three Newmans and their choices represent secularity beautifully, Francis Newman was simultaneously a champion of religious and secular commitments and thus epitomizes in one person the notion of secularity that I use throughout. While 'one of the brightest Oxford undergraduates of his generation'[9] and later a university professor and well known public intellectual, Francis Newman nevertheless has also been largely neglected in Victorian studies, a neglect due partly to the disparagement and dismissals of prominent critics, partly to the prominence of his overshadowing elder brother, and partly to the irrelevance into which some of his religious concerns have fallen.[10] Although the ultimate outcome of his creedal doubts will be familiar to many today, as the prevalence in contemporary parlance of the phrase 'spiritual but not religious' attests, many of Newman's theological preoccupations will appear obsolete to the contemporary reader.[11]

Yet Francis Newman's religious discourse was, somewhat paradoxically, indispensable to the Secularism developed by George Jacob Holyoake at mid-century.[12] Secularism responded to a direct encounter with a particular religious trend commencing at mid-century, of which Francis was the first best representative. Called variously the 'religion of the heart',[13] the 'internal school' of Christian evidences,[14] 'primitive Christianity',[15] 'rational religion',[16] the 'New Reformation',[17] and a kind of English 'mysticism',[18] this tendency held that the guarantee of God's existence was found not in the 'external' Christian evidences; it was not guaranteed by traditional Christian apologetics such as the argument from design or the logic of doctrinal treatises. Instead, the evidence was to be found written within the human heart (or the conscience, feelings, soul, intuition, or instincts). Before some prominent Secularists reconverted to Christianity, justifying their faith in very much the same terms as he did his,[19] Newman's religious prose impacted Secularism proper. Based largely on his personal encounter with Newman and his treatise *The Soul* (1849), Holyoake worked to steer freethought away from dogmatic atheism and toward a new emphasis upon moral sentiments and a positive morality. Such a move would ultimately allow for a compact with theists, including Newman himself, marking Secularism as a distinct freethought movement and creed.

Francis Newman represents a salient example in the history of nineteenth-century secularity for another reason. The chronicle of his spiritual 'progress', *Phases of Faith* (1850), finally persuaded Charles Darwin that 'there was no resting-place *en route* from Anglicanism through Unitarianism to a purely theistic belief'.[20] That such a relatively marginal figure in Victorian studies made a life-changing impression on the towering Darwin demonstrates the radical contingency of history in general and of the secular in particular. But as is clear from his importance for

Holyoake and probably later Secularist reconverts to Christianity, his texts could also have the opposite effect. They also may have moved some freethinkers in the opposite direction – toward conciliation if not reconciliation with religious belief. It is this ambiguity that marks Francis Newman as an important figure for Secularism and secularity – Secularism regarded as a broad tent movement and moral creed encompassing both religious and secular elements, and secularity understood as an overarching condition that envelops both the secular and religious.

I examine two of Francis Newman's religious monographs. The more elevated and arresting is *The Soul: its Sorrows and Aspirations: The Natural History of the Soul, as the True Basis of Theology* (1849), a statement of his religious position after undertaking a theological pilgrimage that left him embracing a kind of minimalist Christianity, if his religious conviction can be called Christianity. Here Newman naturalized the soul as the spiritual organ whose object of sense is God, and divided this religious apparatus from the remainder of a secular 'husk'[21] – to be shorn off and regarded as unnecessary for, or even inimical to, religious conviction. The second is *Phases of Faith: or, Passages from the History of My Creed* (1850), which I treat first. *Phases of Faith* chronicled the doctrinal and personal trials that Newman faced as he passed from Anglican evangelicalism through seemingly every Christian position possible, finally arriving at the mystical theism discussed in *The Soul*. *Phases of Faith* so affected Darwin that he finally abandoned Christianity. Taken together, these two outliers are extraordinary milestones for the representation of secularity within mid-century religious discourse. They are also representative of a religiosity widely divergent from that of the traditionalist and ascetic High Church Tractarianism undertaken by Newman's elder brother. However, I suggest that both moves – Francis Newman's naturalization of the soul and John Henry Newman's Tractarianism – were driven by the same set of circumstances. Although they were quite distinct, the two spiritual crises represented responses to what both figures saw as the crumbling foundations of Protestant Christianity. And here again, we may add the third Newman. The three Newman reconversions (or de-conversions) depended on a response to the establishment evangelicalism that they each abandoned.

Leaving evangelical Protestantism

In a remarkable passage in *An Essay in Aid of a Grammar of Assent* (1870), John Henry Newman wrote: 'Thus, of three Protestants, one becomes a Catholic, a second a Unitarian, and a third an unbeliever: how is this'?[22]

Here, following hundreds of pages of disquisition about the conversion from Protestantism to Catholicism, Newman apparently alluded to his own family, the case of himself and his two brothers. In a passage that suggested that he had paid more attention to his brothers' writing than he may have otherwise acknowledged, John Henry Newman noted that the Catholic (himself), the Unitarian (Francis William), and the unbeliever (Charles Robert) each cherished a particular belief before their (de) conversions away from evangelical Protestantism. According to John Henry Newman, it was the particularity of their initial commitments to Protestantism as such that could explain their ultimate destinations. The first began by assenting to 'the doctrine of our Lord's divinity', which finally led him 'to welcome the Catholic doctrines of the Real Presence and of the Theoticos [Virgin Mary], till his Protestantism fell off from him, and he submitted himself to the Church'.[23] The second began with bibliolatry, 'the principle that Scripture was the rule of faith', coupled with the Protestant principle that 'private judgment was its rule of interpretation'. This led him to find that the Nicene and the Athanasian Creeds 'did not follow by logical necessity from the text of Scripture' and finally to decide that 'The word of God has been made of none effect by the traditions of men'. This in turn led him to end in 'primitive Christianity and to become a Humanitarian'.[24] The third 'subsided into infidelity' because he began with the 'Protestant dogma ... that a priesthood was a corruption of the simplicity of the Gospel'. He rejected the mass, the sacraments, baptismal regeneration, and Christian dogma in turn. Disqualifying all Christian teachers for an unmediated relationship with God, he soon concluded 'that the true and only revelation of God to man is that which is written on the heart'. This passage also would seem to apply to the second Protestant (Francis William), as well. While this creed sufficed temporarily for the third hitherto Protestant, and he remained a Deist for a time, the next logical step was to suggest that the law written on the heart did not require God as a warrantor:

> that this inward moral law was there within the breast, whether there was a God or not, and that it was a roundabout way of enforcing that law, to say that it came from God, and simply unnecessary, considering it carried with it its own sacred and sovereign authority, as our feelings instinctively testified.[25]

Turning now to the physical universe, our third subject finally 'really did not see what scientific proof there was there of the Being of God at all, and it seemed to him as if all things would go on quite as well as at present, without that hypothesis as with it; so he dropped it, and became a *purus, putus* Atheist'.[26] John Henry Newman's illustration suggested that while he had been nominally a Protestant, unlike his two brothers, his private commitment had been to the Catholic and High Church side of Anglicanism. As he wrote in 1840 to his sister Jemima, the point that Newman meant to demonstrate was that '[w]hether or not Anglicanism leads to Rome, so far is clear as day ... Protestantism leads to infidelity'.[27] According to Willey, Newman's ultimate argument amounted to the proposition 'that there is no logical standing-point between Romanism and Atheism'.[28] In fact, in letters to his sister Jemima in 1840, Newman clearly laid out what he saw as the only two choices dawning for the believer: the Roman Catholic Church and infidelity.[29] Yet, as Willey noted, certainly the case of Francis disturbs this inevitability – that is, if we are to deem Francis an honest doubter of creeds and an honest theist.

But what was it about evangelical Protestantism that would lead to such losses of faith and what about it might permit such diverse creedal outcomes as the three Newman brothers evinced? David Hempton suggests in the introduction to his portraits of evangelical disenchantment that although traditional Christianity itself was vulnerable to skepticism especially from the mid-nineteenth-century on, evangelicals might be particularly prone to disenchantment 'because so many are swept into the tradition at a relatively young age, and because the claims and aspirations are so lofty while the liturgical management of failure and dissatisfaction is so weak'.[30] Yet this explanation, and it is by no means the only one that Hempton floats,[31] does not account for the diversity of the beliefs that the disenchanted finally embraced, nor does it serve to explain just what such ultimate commitments were liable to be.

Such questions as these cannot be answered definitively with reference to the impact of the German Higher Criticism, the findings and claims of geological and evolutionary science, the growth of knowledge about other cultures and religious traditions, an increasing religious pluralism, and so on. Such explanations fail to account for reconverts to Catholicism, for example. Yet, it is enough to say that while traditional Christianity at large was subject to new vulnerabilities, evangelicalism was especially susceptible. As Hempton puts it, '[a] tradition built so much on the inspiration and authority of Scripture and on the vital importance of supernatural events such as the Virgin Birth and the physical resurrection of Jesus Christ was especially vulnerable to the new climate of thought'.[32]

In fact, even in the case of John Henry, evangelicalism was the primary target of criticism. In his reconstruction of the early John Henry Newman, that is, the Tractarian Newman seen not from the standpoint of *Apologia pro Vita Sua* (as Turner suggests those caught up in the 'Newman industry' are far too prone to view him) but rather from the perspective of his early Tractarianism itself, Frank M. Turner notes that evangelicalism seemed to invite dissent, and from the perspective of the Tractarians, it was the object of particular repudiation:

> Evangelical Protestantism and not political or secular liberalism in and of itself had been the Tractarian enemy. It was this assault against evangelical Protestantism lying at the heart of the Tractarian Movement that fourteen years later Newman omitted and concealed in the *Apologia*, where he wrote as a controversialist seeking to reshape and rescue his personal public reputation.[33]

Further, Turner sees the conversions or reconversions of Francis and John Henry as parallel even though antithetical. Turner focuses on the commonality of the two Newman trajectories. After all, whatever their final destinations, each had undergone a religious crisis that effectively resulted in the renunciation of established evangelicalism:

> The most fundamental religious experience of Newman's life was his adolescent conversion to evangelical religion. His reception into the Roman Catholic Church almost thirty years later represented the final step in what had been a long process of separation from that adolescent faith. That the conclusion of the process, which commenced in his mid-twenties, was Roman Catholicism does not make it any less a loss of evangelical faith than if, like others of his and later generations, he had ended in Unitarianism, like his brother Francis, or in agnosticism ... If his brother had not more than a decade earlier already used the title to describe his own religious journey, John Henry Newman might just as easily and correctly have entitled the history of his own religious opinions *Phases of Faith*.[34]

Stemming as it did from a reaction to Enlightenment reason,[35] evangelicalism was especially vulnerable to critical rationalist attack. One response was to undertake that attack and another was to seek refuge in a system whose authority depended on deep tradition, apostolic succession, ecclesiastical and ritualistic aestheticism and submission – a kind of sheltering from the attack. The latter is precisely what John Henry would do.

The religious history of John Henry Newman has been amply told and Frank M. Turner has recently undertaken a critical approach to his life and works that seeks to undermine what Turner sees as a long tradition of reconstructed hagiography and its attendant distortion. I will discuss much of the religious history of Francis below. Likewise, I turn now to the case of Charles Robert.

Charles Robert Newman's life history has been pieced together and drawn into several sources such that a complete picture can now be drawn.[36] Strangely, like that of Francis, it crosses paths with George Holyoake and his *Reasoner,* the only publication to which Charles submitted his written thoughts, probably in response to Francis's religious writing. But Charles's religious history begins like that of John Henry and Francis, under the tutelage of the evangelical preacher, Rev. Walter Mayers at the Ealing School in Ealing. By 1821, however, at age nineteen, he already was a votary of Robert Owen, whose social environmentalism he particularly appreciated.[37] Character and belief were products of the social environment; likewise, ascribing culpability based on them amounted to an error. This, Charles believed, applied to himself and thus he would ask others to consider it before judging him. In 1823, John Henry recorded in his journal his discussions on religion with Charles, who by this time had anticipated some of Francis's criticisms of the Bible and the creed of Calvinism. In particular, he found the doctrine of eternal punishment ludicrous, hinging as it did on the meaning of the Greek word, *αἰώνιος,* which Francis would later translate as 'secular'. (See below.) By 1825, John Henry lamented that his brother Charles was probably lost to the faith 'for a long time'.[38]

Yet John Henry felt obliged to help his brother, and 'through the influence of the father of his best friend at Trinity, J. W. Bowden, obtained for Charles a fairly good position as clerk in the Bank of England, where in 1827 he was earning £80 a year'.[39] Here he worked for a few years, until writing offensive letters to the bank directors, who encouraged him to retire. He soon took up a position as an usher at a private school near Hurstmonceaux, Sussex, in the parish of the future Archdeacon Julius Hare, but he experienced problems disciplining the pupils; at one point during an altercation Charles apparently bit a student, ending that short stint. It is unclear whether Charles then married or merely suggested to his family that he was married, but the relationship ended with his wife or fiancée having pawned all of his belongings for alcohol, including his clothes, leaving Charles penniless, wearing nothing but underwear, and lying on a bed of straw, where a friend from the bank found him at No. 7, Hope Place, Bird Street, West Square.

Several positions as school usher followed, and then an aborted attempt to earn a literary degree at Bonn in his early forties. Finally deciding that he was unable to work, Charles sought and received the financial support of John Henry and Francis and moved to the water town of Tenby, Wales, where he lived a Spartan life as a recluse, and where he met his only friend in town, the novelist Thomas Purnell, who was to write his obituary for *The Athenaeum*. It is here that Charles apparently drafted the essays that he submitted to the *Reasoner* for publication, and which Holyoake did publish. These contain arguments against design, a defense of reason, and an argument for the rights of reason. They are dry, terse, and tedious, containing none of the bombast characteristic of earlier freethought rhetoric.

Timothy Larsen has argued in the context of mid-Victorian plebeian radicalism and unbelief that intellectual isolation of the working classes and exclusion from formal education concomitant with an exposure to a welter of popular literature led to a 'suspicious outlook' among the working classes and the belief that the 'higher orders were attempting to dupe them with erroneous theories'. It was this isolation and attendant suspicion that set the stage for unbelief, while 'reconversions [to Christianity] were part of a wider reintegration in Victorian society' later in the century.[40] We can apply this interpretation, *mutatis mutandis*, to Charles Robert. Excluded from the formal education that his brothers received at Oxford, Charles very well might have experienced intellectual isolation and a distrust of established institutions, including the Church. This might explain why, of the three Newmans, only Charles Robert left establishment evangelicalism for outright atheism. And it may also help to explain why Charles never reconverted back to Christianity; he was never able to re-integrate into Victorian society, if he had ever been an integrated part in the first place.

Yet, this explanation does not account for the fact that atheism became an option as such, and we must explain this in terms of the social environment. Unbelief became a means of dissenting from the established order but the question is, why this particular form of dissent, as opposed to another? Of course, we must look to the fact of the established Church and the demands of compliance that it made on those who would advance through an Oxbridge education. Likewise, unbelief or any form of infidelity from orthodoxy became a major stream of dissent, and this stream came to be a well-known option, albeit a painful and difficult one, for the expression of dissatisfaction and rebellion. The outcomes of these and other such crises of faith have been termed 'secularization'. But as we shall see in the case of Francis Newman, the options were not only those of unbelief or orthodoxy.

Phases of Faith: a crisis of creed

Near the end of his discussion of Francis Newman's *Phases of Faith* in *The History of Rationalism in the Nineteenth Century* (1906), Alfred William Benn argued for the importance of Francis Newman for a proper appreciation of the Victorian period's secularity. An avowed rationalist himself, Benn described the nineteenth century, beginning with the mid-century decline of evangelicalism, as a period of progressive religious 'dissolution' tending toward secularization. According to Benn, with the encroachment of rational inquiry, especially the direct confrontation with the higher critical Biblical exegeses and the subsequent doctrinal difficulties, the religious edifices were everywhere crumbling. Benn considered *Phases of Faith* emblematic of a tottering Christianity with Newman at the vanguard leading his generation and those to follow along the path of rational liberation:

> I have given a somewhat extended analysis of Francis Newman's arguments, partly because they constitute the most formidable direct attack ever made against Christianity in England, and partly because of their immense historical importance, published, as they were, at a critical period in the intellectual life of the nation. When 'Phases of Faith' appeared, discontent was simmering in all directions, but no controversialist had as yet come forward to canvas the popular creed point by point, and to reject all the most prominent dogmas that brought it into collision with the new physical science, the new historical criticism, and the moral principles which, though not new, had been temporarily darkened by the pietistic revival [evangelicalism]. Carlyle had not cared, Grote and Mill had not dared to publish their opinion of the reigning religion; Charles Hennell had spoken without the authority of a scholar. Francis Newman was a scholar armed at all points, whose competence none could deny; and not only a scholar, but a master of clear and impressive language, the apt vehicle for a masculine, straightforward logic which puts the tortuous sophistry of his brother to shame.[41]

Similarly, Basil Willey characterized Newman as a pioneer beating down the once-treacherous mountain pathways leading to the safety of secular vantage points, from which one could overlook a recently gained secular territory:

> His difficulties may have ceased to terrify, and his conclusions – many of them – are now the unquestioned data of every beginner. Yet it remains an interesting and touching spectacle to watch a pioneer winning, inch by painful inch, the vantage-ground we now occupy.[42]

In *The Newman Brothers* (1966), William Robbins characterized Francis as a trailblazer who 'made his contribution to the secular shape of things to come'.[43] In addition, more recently, Kathleen Manwaring placed Newman 'at the forefront of the secular humanism of the twentieth century'.[44]

Although the paeans for Newman's secularism and rationalism slowly accumulated over the twentieth century, even Benn, who referred to Francis Newman as a 'champion of rationalism', admitted that Newman was not himself a rationalist. Instead, he was a devout religionist of a particular stripe:

> I have called the younger Newman a champion of rationalism. And in point of fact he was forced into that position by circumstances; but I must not be understood to imply that he was a rationalist in the complete sense, or indeed in any sense that would give reason a preponderance over faith. With him, as with his brother, the dominant trait was *a morbidly introspective mysticism* allied in both with the keenest dialectical ability.[45]

In personal correspondence in 1860, Newman disclaimed the moniker of rationalism, suggesting instead, that 'if I am prominent in anything, it is not a Rationalism ... but it is by my spiritual writings ... My doctrine is Spiritual Theism, as opposed to Atheism, Pantheism or Old Deism or German Rationalism'.[46]

In connection with Francis Newman and in accord with the recent challenges to the standard secularization thesis, W. F. Bynum notes 'the persistent hold of religion throughout the nineteenth century, even on those for whom doctrine seemed irrelevant'.[47] Bynum reminds us that 'giving up one's faith rarely happened without a struggle, and movement was often *between* established creeds rather than the abandonment of some purpose or plan in the universe'.[48] While Newman's crisis did not involve a movement between two *established* creeds, it was nevertheless a movement between creeds, the latter creed being one very much of his own making.

How then can we comprehend Francis Newman's peculiar religiosity? How and why did it emerge out of evangelicalism? What was the relationship between his supposed rationalism and his mystic theism? And how do we figure Newman in terms of the emergence of the secular? Does his religious writing represent the dawning of a broad secularism, or merely a second conversion within religious creedal boundaries? Was *Phases of Faith* the account of a progressive and seemingly inevitable secularization, as suggested by Benn, Willey, Robbins, and Manwaring? Or, as Newman himself suggested, was it rather the means by which the secular – having already invaded religion – was ultimately arrested and a true spirituality, a 'primitive Christianity', recovered?

Certainly, *Phases of Faith* may be understood as representing Newman's disenchantment with traditional, evangelical Christianity. On the other hand, *The Soul* represents the continued enchantment or re-enchantment of a particular religiosity, what he deemed a wholly experiential religiosity. This interpretation takes into consideration Newman's use of rationality to decompose his creed in *Phases of Faith* while also acknowledging his enduring faith commitment as evidenced in *The Soul*. With this dual movement of disenchantment and enchantment/re-enchantment, Newman finally arrived at the basis of true religion as he saw it, a conviction that legitimate spirituality has little or nothing to do with espoused creeds and everything to do with the individual relationship of the soul with God. As such, Newman's crisis was not properly a 'crisis of faith' but rather a 'crisis of *creed*'. Alfred, Lord Tennyson's famous lines in *In Memoriam* represented an apt restatement of the position Newman articulated in *Phases of Faith*: 'There lives more faith in honest doubt,/ Believe me, than in half the creeds'.[49] Although his critical supporters in the twentieth century believed that he had stopped just short of the inevitable unbelief and secularism of others to follow, Newman actually maintained that he had deepened his religious commitment to what he took to be the only legitimate and fundamentally sound position possible for a critical seeker after the truth in the mid-nineteenth century.

As Newman maintained in the Preface, Newman did not intend *Phases of Faith* to be a an autobiography; it was not his (earlier) version of John Henry's *Apologia pro Vita Sua*. The motives for the book, Newman explains at the outset, were from '[p]ersonal reasons the writer cannot wholly disown, for desiring to explain himself to more than a few, who on religious grounds are unjustly alienated from him'.[50] Written in the form of a confession, *Phases of Faith* was meant to exonerate Newman from blame, to vindicate himself before family and friends whose trust and friendship he had lost or endangered, to secure his innocence and righteousness for adopting the positions that he had held and for finally rejecting traditional Christian doctrine, Biblical inerrancy, and historical Christianity. Organized into chronological periods or phases, which as David Hempton has noted reflected the periodic dispensationalist evangelicalism of his youth,[51] *Phases of Faith* is a prosaic recounting of Newman's creedal dissolution and disillusionment but also it is an impassioned plea for a new kind of religiosity. As U. C. Knoeplmacher pointed out, Newman was 'not concerned with endearing himself to his readers, but rather wants to serve truth by conducting a self-examination that is as rigorous and objective as the introspective assessment formerly demanded by Evangelical discipline'.[52] Meanwhile, as Manwaring notes, according to Francis's *Athenaeum* obituary, *The Soul* and *Phases of Faith* were sent to his brother, who was by then a Roman Catholic, but Francis 'bitterly declared that his brother said he had not time to read them'.[53] Whether or not this is the case, as we have seen, John Henry had a fair idea of what Francis's *Phases of Faith* and *The Soul* entailed.

Phases of Faith begins with Francis Newman at approximately age fourteen. Upon exposure to the persuasive Anglican evangelical minister Rev. Walter Mayers at the Ealing School, Newman began his religious quest with an 'unhesitating unconditional acceptance of whatever is found in the Bible'.[54] Later, Newman served as an assistant at Mayers's parish, where John Henry would preach his first sermon.[55] Soon after arriving at Oxford, however, Francis could no longer reconcile either the Thirty-nine Articles or Sabbatarianism with the letter of Scripture. He hesitated before subscribing to the articles upon graduation from Oxford given his disagreement with the article on infant baptism, but decided 'that it had no possible practical meaning to me, since I could not be called on to baptize, nor to give a child for baptism'[56] – although he soon came to regret this compromise of principles. After graduating from Oxford in 1826 (with an extremely rare double first in math and classics; John Henry had won no firsts), he became a Balliol fellow. Upon accepting his degree, 'the whole assembly rose to welcome him, an honour paid previously only to Sir Robert Peel on taking his double first'.[57] In 1827, based on his rejection of infant baptism, Article Twenty-seven of the Thirty-nine Articles, he gave up his fellowship and thus the prospect of an MA and the life of an Oxford scholar. He consequently abandoned his original intention of taking orders. Instead, he traveled to Ireland to become a tutor in the household of Edward Pennefather, who later became the chief justice of Ireland.[58] In the course of eighteen months' residence, he made the acquaintance of the Irish clergyman John Nelson Darby, Pennefather's brother-in-law and leader of the fledgling secessionist pre-millennial evangelical Plymouth Brethren. Falling deeply under Darby's spell, Newman attempted to emulate Darby in piety and pre-millennial, anti-worldly self-renunciation and anti-intellectualism.[59] As we shall see, it was the last of these that made Darby's influence difficult for Francis to bear. When he returned to Oxford, Newman renewed his friendship with Benjamin Wiles Newton, a Calvinist evangelical whom he had tutored in Exeter. Later, he introduced Newton to Darby, and the two became significant figures in the early days of the Plymouth Brethren, although they later went separate ways.[60] During his stay in Ireland, Newman also became acquainted with Anthony Norris Groves, another of Darby's disciples. Newman read with admiration Groves's pamphlet *Christian Devotedness* (1829). It called for self-renunciation, willful poverty, and a singular devotion to faith in anticipation of the imminent return of Christ. Under the influence of Darby and Groves, Newman was a premillennial evangelical anticipating the Second Coming of Christ.

It was with this creedal commitment that Newman joined a missionary troupe of like-minded evangelicals from the fledgling Plymouth Brethren fold for an ill-fated mission to the Middle East. These brethren included John Vessey Parnell, Parnell's son and the latter's fiancée, Nancy Cronin, her recently widowed brother (a medical doctor) and his infant daughter, as well as the Cronins' mother, and a Mr. Hamilton. The missionaries left Dublin on 18 September 1830 with the intention of traveling directly to Baghdad, where they were to join Groves and his wife, who were already 'busy converting souls'.[61] Newman's cultural experience in Palestine and Persia would leave a lasting impression upon him regarding the likely impotence of the Christian Scripture and doctrine for converting Muslims given the embedding of religion in culture.[62] Newman provided an example of his disenchanting attempts at converting a Muslim carpenter at Aleppo, who answered his proselytizing attempts as follows:

> God has given to you English a great many good gifts. You make fine ships, and sharp penknives, and good cloth and cottons; and you have rich nobles and brave soldiers; and you write and print many learned books (dictionaries and grammars): all this is of God. But there is one thing that God has withheld from you, and has revealed to us; and that is, the knowledge of the true religion, by which one may be saved.[63]

Meanwhile, upon returning to Britain after this disastrous missionary experience, his apostolic confidence deeply shaken, he remained for a time under the tutelage of Darby, who instructed him to abandon all creeds and turn to the Bible alone. Newman continued his extensive study of the New Testament and soon became convinced that whatever the apostles John and Paul thought about Christ's divinity, they did not teach the doctrine of the Trinity as such: 'in studying this word; I found John and Paul to declare the Father, and not the Trinity, to be the One God'.[64] His own interpretation of the New Testament on this key point brought him into direct conflict with Darby himself, who declared that the Father was the Trinity. He discussed the subject with other missionary friends and found – owing to letters written by Darby forewarning them against 'sheltering' such an apostate – that he was avoided and denounced as both immoral and a heretic, merely for undertaking the very kind of direct study of Scripture that Darby had recommended.[65] Rather than independently interpreting the Bible as claimed, as Newman saw it, Darby had actually filtered his beliefs through a pre-existing doctrine, in this case Trinitarianism. Newman felt utterly betrayed. At nearly the same time, hearing that Francis had upon occasion delivered addresses to small private religious gatherings, his brother John Henry felt duty-bound to give up all contact with one who presumed to undertake the duties of the priest. As his elder brother then supported the Newman family, the ban effectively extended to the rest of his family members, including his mother and three sisters. Thus, cut off from the support of family and friends, with even new acquaintances having been turned against him, Francis exclaimed:

> My heart was ready to break: I wished for a woman's soul, that I might weep in floods. Oh Dogma! Dogma! How dost thou trample under foot love, truth, conscience, justice! Was ever a Moloch worse than thou? burn me at the stake; then Christ will receive me, and saints beyond the grave will love me, though the saints here know me not. But now I am alone in the world: I can trust no one.[66]

Already treated as a pariah for a single (albeit central) article of 'honest doubt', he came to distrust the morality of the Christians who had rejected him. As Benn would put it, '[a]ltogether, pious people behaved so badly as to convince the young man that there is no necessary connexion between religion and morality'.[67] Thus, he reasoned, morality must be based on something more primitive than doctrinal convictions or even Biblical sources. In fact, following dogma or a sacred text could inveigh against moral behavior: 'They have misinterpreted that word: true: but this very thing shows, that one may go wrong by trusting one's power of interpreting the book, rather than trusting one's common sense to judge without the book'.[68] Deeply shaken from his missionary jaunt and now unsettled in his creed and enduring a shaming on that account, Francis concluded that the true end of religion was morality and religion a mere means to that end.[69] Thus, he came to believe in the independence of morality from religious indoctrination, a point that would greatly impress Holyoake.[70] But it marked him as already errant in terms of traditional Christianity.

Having arrived at the independence of morality, soon parts of the New Testament failed to meet his moral standards. He was no longer merely comparing doctrinal statements with Biblical sources and finding the prior wanting or errant; he held the Bible itself up to scrutiny. As such, with a sharp axe of rationality, he chopped into the trunk of Protestant Christianity itself. In particular, in the third period recounted in his theological opus, 'Calvinism Abandoned as Neither Evangelical Nor True', he elaborated upon the difficulty of squaring the Biblical notion of eternal reprobation with his own sense of moral equity. Having once stealthily peeked into a Unitarian treatise about the doctrine of eternal punishment, the standard arguments against the traditional notion of hell came easily to him. First, he noted that the word for 'eternal' in the Bible was the Greek, *αἰώνιος*, the equivalent of 'secular': 'belonging to the ages' or signifying 'distant time'. Thus, 'eternal' damnation might indicate at most a long-lasting punishment or punishment carried out at a future time.[71] He could reconcile this interpretation with his own moral intuition that a finite being could not or should not endure infinite punishment for finite transgressions. If an eternity of punishment awaited the damned, then the atonement of Christ amounted to a failure, with Satan and sin the final victors carrying off the vast majority of souls. It would have been better had the human race never been created in the first place.[72]

I will spare the reader Newman's torturous disquisitions on the Nicene and Athanasian Creeds. Christology engaged him for some time; he deliberated the positions of the Athanasians, the Arians, the semi-Arians, and the Sabellians, resting briefly at semi-Arianism or a belief in Christ's divine yet created nature rather than his consubstantiality with God the Father. The issue of Christ's nature became important in consideration of the atonement; *how could God die by crucifixion*? The difficulties of imagining the death of God led him to adopt full Arianism, or a disbelief in the divinity of Jesus Christ. Finally, the inability to accept the doctrine of 'The Fall of Man' and the corruption of all humanity by Adam's sin meant that he had broken from Calvinism entirely.

In the fourth period, entitled 'The Religion of the Letter Renounced', Newman brought his reading of the German Higher Criticism and the findings of geology to bear, subjecting both Old and New Testaments to a ruthless historical, moral, and scientific analysis. Suddenly, as if reading the Scriptures for the first time, he found errors of historical fact, natural impossibilities, specious miracles, grave immorality apparently accorded approbation, and even egregious misquotes and misinterpretations of the Old Testament by the New Testament writers. Refusing to impute miracles for every naturally implausible event in the Old Testament and with recent findings in geology (namely, Lyell's *Principles*; see Chapter 2) indicating the impossibility of a universal deluge as well as evidence of the existence and death of animals and plants long before the introduction of humankind,[73] he could no longer maintain belief in the inerrancy of the Bible.[74] Further, the Pentateuch was obviously not written by a single person (Moses); it was apparently a shoddy patchwork of multiple sources accumulated across various periods.[75] He found poetic figuration converted into miracles by later writers.[76] Whole books of the Bible began to fall away not only for lack of historical accuracy but also for lack of divine inspiration.[77]

In passages that echoed Wesley's dismissal of 'opinions',[78] the conclusion he drew from this period was that 'true apostles' did not argue over creeds and '[s]entiment surely, not *opinion*, is the bond of the Spirit; and as the love of God, so the love of truth is a high and sacred sentiment, in comparison to which our creeds are mean'.[79] Newman quoted Samuel Taylor Coleridge: 'If any one begins by loving Christianity more than the truth, he will proceed to love his Church more than Christianity, and will end by loving his own opinions better than either'.[80] Like morality, truth not only preceded but also superseded creedal commitments and the Bible. The truth of God's existence was to be discerned in a direct relationship with the divine and was neither dictated by a Church nor garnered from the Bible. Scholars should adjudicate other truth claims – those relating to history and the physical sciences, and so on.

On the one hand, Newman suggested that 'external' doctrines might be a positive obstruction to true spirituality. Stringent creeds often constrained the spirituality of the believer and limited her effectiveness at doing good works. On the other, he suggested that doctrines were of little or no consequence and that his own shifting 'belief' or creed – which he distinguished from faith – had not altered his spiritual condition. In a trope reminiscent of Carlyle, the 'internal' represented true spirituality, while the 'external' (doctrines, traditions, creeds) were superfluous or even obstructive to it. Judging others according to their creeds was thus a spiritual and moral error and he lamented having done so himself, especially in the case of his elder brother.[81]

> I had a brother, with whose name all England was resounding for praise or blame; from his sympathies, through pure hatred of Popery, I had long since turned away, What was this but to judge him by his creed? True, his whole theory was nothing but Romanism transferred to England: but what then? I had studied with the deepest interest Mrs. Schimmelpenninck's account of the Portroyalists, and though I was aware that she exhibits only the bright side of her subject, yet the absolute excellencies of her nuns and priests showed that Romanism as such was not fatal to spirituality.[82]

Thus, he wrote John Henry to ask for his pardon and received a warm reply. Still Francis could not understand 'how a mind can claim its freedom in order to establish bondage'.[83] His acute and chronic disdain for Romanism or Popery was evident on nearly every page.

By the end of his fifth and sixth periods, Newman had renounced 'second-hand faith' and declared that 'history is not religion'. By second-hand faith he meant the presentment of miracles as an inducement to faith; faith based on the witnesses of the apostles, including all the witnesses of the resurrection; the Old Testament prophecies, especially those supposedly predicting the arrival of Jesus Christ as the Messiah; the New Testament prophecies, including those purportedly spoken by Jesus; and the supernatural origin of Christianity itself.[84] Because very few persons had the necessary time and scholarly training to discern the historical veracity or determine the literary import and meaning of Biblical texts, faith as such could not be deduced from Biblical evidence or based on the testimony of others. Instead, *'the moral and spiritual sense is the only religious faculty of the poor man'*.[85] By declaring that 'history is not religion', Newman meant that the Bible could not be understood as valid history, that faith did not depend upon the veracity of the accounts told in it and that the Christian narrative, especially the 'Messiahship' of Jesus Christ, was not an essential kernel of faith. He could find no justification for it in the Scriptures. Further, the myth of Messiahship was the root of all evil in Christianity, teaching as it did that the present life is merely a waiting room for another, thus encouraging lack of attention to the present conditions. His Protestantism had constrained his sympathies for others and selfishly restricted his concerns to his own salvation, thus curtailing his good works.

What then, if anything, was left of his Christianity, he wondered?[86] He had denied the divine inspiration of the Bible, ridiculed the histories in both Old and New Testaments, doubted the veracity of the Christian miracles and devalued their importance for faith, dismissed the narrative of salvation through the death and resurrection of Jesus Christ, and abandoned the anticipation of the Second Coming as a real eschatological eventuality. Yet, he continued to 'love and have pleasure in so much that [he] certainly disbelieved'. He found satisfaction in reading Christian Scripture and other Christian literature, including Paul, Luke, and 'the verses of some hymns'. Although he found grave errors in them, some Christian texts, especially Paul, still brought him enjoyment and satisfaction. He wondered why. The answer would define the meaning and truth of religion and spirituality for Newman: 'my religion always had been, and still was, *a state of sentiment* toward God, far less dependent on articles of a creed, than once I had unhesitatingly believed. The Bible is pervaded by a sentiment, which is implied everywhere, – viz. *the intimate sympathy of the Pure and Perfect God with the heart of each faithful worshipper*'.[87]

Clearly, Newman had turned evangelicalism's methods on evangelicalism itself by emphasizing the importance of 'internal' dispositions over systematic theology. Further, echoing and altering Feuerbach, Newman continued, the 'truest essence and most vital point in Christianity' was to draw fructification from the evidences of divine communion represented in the lives and writing of 'great souls', and to join in their joys, sorrows, and hopes.[88] That is, 'Christianity' for Newman amounted to the testimonies of the evolving relationship between 'God and Man', as found in the writings and teachings of great souls. Jesus was an imperfect, misleading, and understandably misrepresented teacher. No clear record of his life or work existed and his divinity and Messiahship were fables. His mission had been to break the grip of the dead letter on Judaism and to establish a community of equality in faith and love.

Newman concluded the first edition of *Phases of Faith* with an appeal for religious progress and against religious bigotry. While known for constraining the sympathies and increasing the hatred among believers for outsiders, the worst evil religious bigotry wrought was the reaction against religion itself so that piety was sacrificed as 'men are lapsing into Atheism and Pantheism'.[89] In a passage that he apparently directed at the then current freethought movement and that possibly indicted his brother Charles, Newman suggested that atheism was the first mistaken response to religious bigotry, undertaken by a 'class of persons [that] inveighs warmly, bitterly, rudely against the bigotry of Christians'. They not only underestimated the depth of feeling within the Christian community and thus the intransigence of religiosity but also they served to reify and consolidate that which they attempted to extirpate: 'Hence their invective is harsh, cold, unsympathizing; and appears so essentially unjust and so ignorant, as to exasperate and increase the very bigotry which it attacks'.[90] The second mistaken response to bigotry was silence, of which the clergy themselves were guilty. Their reticence had the effect of encouraging the adhesion to an outworn creed. Much as Carlyle had written in *Sartor Resartus*, Newman argued that only by continuously reconciling inner truth with outward forms could religion survive and progress. Progress would come by way of the pruning of credulity and superstition by skepticism. Otherwise, the critics remained banished from the flock and 'despise instinctive religion' and religious persons remained dreadful of 'critical and searching thought'.[91] Newman's final plea – which would be later echoed by reconverts to Christianity from Secularism[92] – was for the incorporation of freethought within religion itself so that religion might be as impartial as modern science with respect to the pursuit of truth:

> The fault lies undoubtedly in the fact that Practical Devoutness and Free Thought stand apart in unnatural schism. But surely the age is ripe for something better; – for a religion which shall combine the tenderness, humility and disinterestedness, that are the glory of the purest Christianity, with that activity of intellect, untiring pursuit of truth, and strict adherence to impartial principle, which the schools of modem science embody.[93]

Without such incorporation of criticism within religion as such, Christianity would be condemned to a battle with advancing scientific, historical, and other knowledge, which would endanger its very survival.

Although he did draw on Strauss and De Wette for some insights, Newman had come to most of his conclusions without the help of the German Higher Criticism. (He had also drawn on the direct assistance of British friends and religious scholars, including Thomas Arnold.) Unlike the Higher Criticism, however, *Phases of Faith* was an intimately personal testament of the struggle with a hitherto espoused doctrine and its unraveling, layer by layer, before the emboldened and embroiled critical faculties. His own moral rejection of Biblical ethics had preceded and was in fact more important to his de-conversion from evangelicalism than his reading of the Higher Criticism or his encounters with geological and evolutionary science.[94] These merely served to confirm his own creedal doubts and to affirm his commitment to a faith without dependencies on traditional Christian doctrine. Thus, *Phases of Faith* inaugurated a new genre of confessional religious literature of doubt in the period; it would be followed shortly thereafter by James A. Froude's *Nemesis of Faith* (1851), a novelistic account of a 'crisis of faith' that Francis found far too generous to John Henry.[95]

Critical reception of *Phases of Faith* was nearly immediate and often quite bitter. *The British Quarterly Review* ridiculed its author, mocking his supposed pretensions to originality, lambasting his reasoning and finally dubbing the book 'one of the most dishonest in the English language'. The reviewer ended by charging Newman with a 'deliberate purpose to mislead'.[96] *The North British Review* soon followed, only this reviewer's tone was 'mournful' and blamed the educational system at Oxford, not only for Francis's 'declension from something like an evangelical profession to the gulf of utter unbelief' but also for John Henry's having 'passed over into the Church of Rome', after 'having exhausted the ritualism of the Church of England'.[97] *The Journal of Sacred Literature* found Newman naïve and never in possession of a firm foundation in Christian doctrine or belief from the outset.[98] The Bostonian *Brownson's Quarterly Review* noted Francis's 'respectable scholarship', 'kind and warm heart', and 'honesty and sincerity', but then expressed 'regret to see him wasting his fine powers and attainments in the unpraiseworthy effort to obliterate faith from the human heart, and reduce mankind in their estimation to a level with the beasts that perish'.[99]

Pamphlet- and book-length denunciations followed, including David Walther's *Some Reply to 'Phases of Faith'* (1851), John Darby's *The Irrationalism of Infidelity* (1853), Henry Rogers's novelistic *The Eclipse of Faith; Or, A Visit to a Religious Sceptic* (1852) and *A Defence of 'The Eclipse of Faith'* (1854). Walther's pamphlet aimed to 'fore-arm *Christian readers* against an attack upon their faith',[100] while *The Irrationalism of Infidelity* was an anonymously published and lengthy defense of Christianity and a scathing attack on *Phases of Faith*, which supposedly intended to spare the author:

> To the book, I can measure out, without a pang, unmingled feelings of disgust and contempt. To the author I could not. The thought of him awakens sorrow, regret, pain, a thousand feelings which the evil I find in his work, the thoughts as to Christ once expressed by him and supposed genuine by me, and my own love to souls however feeble – as alas! it is – contribute to produce.[101]

Darby indeed meted out disgust and contempt for the book, while also making imputations about the author's character, sincerity, and reasoning abilities along the way.

Rogers's *The Eclipse of Faith* is an epistolary novel set in mid-nineteenth-century. Its characters discuss the religious controversies in an England under altered religious conditions. The Tractarian movement led by John Henry Newman has succeeded in Romanizing the Anglican Church, whose services no one attends, and the religious school inaugurated by Francis Newman has acquired its share of followers. Meanwhile, a general religious skepticism festers and grows. George Fellowes is a Newman acolyte whose new spirituality Harrington, the novel's skeptic, treats with incredulity and barely concealed derision. Harrington interrogates the Newman disciple, exposing contradictions in Newman's claim for the impossibility of 'external' or 'book-revelation' and the possibility of 'internal' revelation only. If Newman's book had been necessary to reveal the truth of 'internal' revelation to Fellowes, how then was this truth 'internal'? Why was the book necessary to bring out this internally revealed truth? If such a truth had been revealed internally to all, then Newman's book would have been superfluous at best and impertinent at worst. Then again, why were only some people privy to such an internal revelation, while others groped in the dark? Was the difference not something 'external' after all? Why was the revelation in Newman's book admissible but that of the Bible inadmissible? Further, Harrington quizzed Fellowes on Newman's conception of spiritual progress, which appeared to contradict the absolute truth of divinity residing internally. But the major objections of the *Eclipse of Faith* were directed mostly at *The Soul*, to which I will now turn.

The naturalization of *The Soul* and the spiritualization of evolution

As the title suggests, Francis Newman's *The Soul: its Sorrows and Aspirations: The Natural History of the Soul, as the True Basis of Theology* (1849) is a paradoxical text. While clearly alluding to Robert Chambers's anonymously published *Vestiges of the Natural History of Creation* (1844), a book that became a 'Victorian sensation' just five years earlier,[102] Newman raised the stakes of the naturalizing tendency of the period by proposing to naturalize the soul itself and treat its study much as one would that of a physiological organ. At the same time, however, he discussed the soul as that organ that shares in divinity, the organ whose function is the perception of the infinite and whose province is an intimately personal relationship with the Divine Creator. On the one hand, we can see the book as a precursor to William James's *The Varieties of Religious Experience* (1902), as it claimed to launch a study of the soul's ontogeny and pathologies as well as undertaking a psychological treatment of religious experience. Newman conceived of his project as inaugurating 'a science of God', which for him was the proper definition of theology.[103] On the other hand, despite offering such a rationalized explanation for spirituality, in *The Soul* Newman 'occasionally reaches heights of almost mystical intensity',[104] recalling even the ecstatic contemplations of such Catholic saints as Saint John of the Cross and Theresa of Ávila. Finally, the treatise ends by offering yet a third dimension: a diagnoses of the state of Western European Christianity, a prescription for a new kind of religiosity, as well as a spiritual prognosis should this recommendation go unheeded.

In the first chapter, Newman described the soul as follows:

> In the English tongue, indeed, the very word Soul appears to have been intended to express that side of our nature, by which we are in contact with the Infinite. The Soul is to things spiritual, what the Conscience is to things moral; each is the seat of feeling, *and thereby the organ of specific information to us*, respecting its own subject.[105]

During a period of increasing specialization and the division of labor within science and industry, Newman descried a division of labor within the human being, distinguishing the soul from other 'organs' – such as the conscience, the intellect, and the will. He began by isolating the soul as an organ with a particular set of functions and setting out on a study of its development both in the individual and the species. Drawing parallels to evolutionary theory, he described the soul as an organ with a natural, developmental history. Like the conscience, the soul was not created in a mature state but rather develops historically both ontogenically and phylogenically. It is naturally underdeveloped in children and 'savages' and must be exposed to various conditions or experiences in order to reach maturity. Yet as 'a higher organ' than the conscience, its history is more complex 'and its diseases also are more hidden and more embarrassing, and in consequence its pathology will assume an apparently disproportionate part of a true theology'.[106]

Thus spiritualizing mid-century evolutionary theory, Newman sketched the experiential conditions necessary for the soul's proper development. He examined the means by which the infinite comes to be experienced and 'known' as such. These conditions were essentially environmental, like those acting upon the fetus in Chambers's evolutionary schema. The encounters discussed are treated as a natural progression toward the ultimate objective of a personal relationship with the infinite God and include the appreciation of awe, wonder, admiration, order, design, goodness, wisdom, and reverence. Each of these senses, which are also periods of development, can have either a healthy, proper outcome, or a diseased one, depending on the personal characteristics and cultural conditions into which the soul is placed. A sense of awe, for example, can be elicited by darkness in such places as groves. The dark produces a sense of the unknown, which is experienced as the infinite. Newman's argument then recalled Auguste Comte, but of course he took the matter in a different direction: one of the 'numberless deviations' that may take place upon the experience of awe is *fetishism*, by which one 'ascribes divine virtue to some common object; to a stone, a beast, a tree, or a scrap of writing'.[107] But fetishism is not confined to 'primitive' cultures; it is also found 'in the midst of enlightened science and highly literate ages' and is especially seen in cases where individuals or groups exhibit 'a positive dread of clear notions' and apparently manufacture mystery 'as if there were danger lest the human mind should exhaust the mysteries of the universe, and leave no room for wonder and reverence'.[108] The primary target for this criticism was Catholicism, but it applied equally to other theological systems, such as any Trinitarianism or other obfuscating doctrine. Such fetishes as the belief in the transubstantiation of the Eucharistic host or the power of priests to release souls from purgatory with special prayers, for example, are as debilitating to the soul as any other primitive talisman and thus retard its development.

But passing successfully through awe without becoming fixated on fetishes, the soul next encounters wonder. In a passage that recalls Romantic reconfigurations of religiosity in general and Edmund Burke's *A Philosophical Enquiry into the Origin of Our Ideas of the Sublime and Beautiful* (1757) in particular, Newman referred to the experience of the sublime as necessary for approaching religious experience:

> There is indeed an elementary religion, a certain religiosity, implied in the perception and enjoyment of the Sublime. The soul, awakened to a sense of the boundlessness of the universe, of its own essential littleness and inferiority, combines an aspiration after fuller knowledge with a devotional self-prostration in the presence of that power, principle, or person, out of which we and all that we see has proceeded.[109]

Again, the experience of the sublime can be perverted and turned into 'mere play with the ideas of things infinite, [which] appears to be more fatal to religion than any other corruption'.[110]

Yet overcoming such potential fatalities, the admiration of beauty and recognition of order in the universe further prompt a sense of the infinite and bring to mind the deity, while a sense of design is suggestive of a designer. The recognition of order, Newman argued, in a passage again reminiscent of Comte, ends polytheism, as the wills of many gods cannot possibly be harmonized in an orderly universe without the superintendence of a single, supreme deity. The deviation associated with the sense of order is atheism, because regularity is suggestive of an absence of mind and will: 'Everything appears either to remain as it is or to change, by a Law. This suggests the theory, that Mind is not in the universe at large, since it is not wanted to account for Motion'.[111] Newman reasoned that those who believe that they themselves have no will but rather that their behavior is determined by causal antecedents naturally believe that the universe is also absent of a will; thus they become atheists: 'for, discerning no first principle of movement even within himself, he of course needs none out of himself'. Yet, the error is not self-consistent, because a person who denies that he has a will would not thereby concede that he has no mind. Thus, he cannot deny the universe a mind merely because he sees no will at work within it: 'Grant that the human mind has no Will; suppose that the divine mind is herein similar: still, that is no reason for denying that there is mind in the universe, in the only sense in which he has experience of mind. If he admits this, he will really become a Theist'.[112] This was one of Newman's refutations of atheism, which he based on extrapolation from the human constitution to the universe at large. The other refutation he made was a from a moral perspective: 'When Atheism depends on the Moral error of believing that man's Will is never self-moving, it is to the Moralist that we must appeal for correction'.[113] This correction involved showing how belief in an absence of will necessarily made morality impossible because one without a will could not act from duty or virtue but only from necessity. The correction of the first error was based partly on imputing a mind to the universe given the experience of the human mind but also it depended on the sense of design in the universe, to which he turned next. Newman argued that design could not be deduced syllogistically but rather must be inferred through intuition or common sense. Further, he suggested that a lack of the sense of design signified a deficiency in the observer.[114]

Once the soul ascertains the existence of one, personal God, it necessarily accepts, *a priori*, the absolute goodness of God. Yet again, in another sense, this goodness is imputed to God based upon the existence of goodness in human beings, and of course, it must be immeasurably superior to the latter:

> To conceive of God at all, as an intelligent existence, and not regard Him as morally more perfect than man, is obviously absurd. Nor only so; but to volunteer limiting any of His attributes is equally absurd. Until the contrary be proved, we unhesitatingly attribute to Him boundlessness in every kind of which we can conceive. But on account of the last limitation, the Perfections of God are justly called a projected image of our own highest conceptions.[115]

With this last sentence, Newman evinced his exposure to the central idea in Ludwig Feuerbach's *The Essence of Christianity*, written in 1841 but not translated into English by Marian Evans (George Eliot) until 1854. No earlier evidence for Newman having read Feuerbach in the original German is extant, but this passage is unmistakably Feuerbachian. Of course, Newman reversed the implication of Feuerbach's argument, which was to suggest that the human conception of God was a projection of our own (potential) goodness and perfection onto an imaginary being. Humanity needed to reclaim the creation for ourselves and thus become what we imagined God to be. Newman, on the other hand, imputed the existence of God from the projection. He thus showed that the Feuerbach's meaning could be reversed just as the process of secularization could be reversed.

The deviation that can occur at this stage is idolatry, which is to worship as infinite that which the finite soul knows to be less than infinite. Since the one who holds an ideal of the infinite is finite, he may have a conception of the infinite that may indeed fall short of it. Yet, the word idolatry should be reserved for cases in which one worships what he knows to be less than his ideal.[116] Here Newman took the opportunity to equate Paganism and forms of Christianity that demand, by means of authority, worship of that which is beneath the ideal of the worshipper. These forms include the 'Bibliolatry' of Protestant and the 'Ecclesiolatry' of the Roman Catholic, the worship of the Bible and of Church authority, respectively. Thus, Newman deemed Catholicism and Protestantism inferior forms of religious worship and placed them on an evolutionary scale leading to 'a higher and purer' form, which apparently he had himself attained.[117]

In the final chapters of *The Soul*, Newman turned to treat a personal relation of the soul to God, spiritual progress, hopes concerning a future life, and the prospects of Christianity. It is in the chapter entitled 'Sense of Personal Relations to God' that Newman's rhetoric soars to mystical heights and leaves off, for the most part, polemics against traditional Christianity. But before this, in answer to some objections to earlier editions, he took on those who suggested that the sense of the infinite and a relationship with God may be a mere delusion on the part of the subject who experiences it, that there might be no objective correlative to the organ of the soul. The soul may merely be experiencing its own feelings, with no real object (God) actually corresponding to said feelings. To this Newman ironically drew on a radical Humean skepticism to suggest that the same may be the case for the object of *any* sense:

> *Perhaps* there is no outer world, and our internal sensations are the universe! Syllogistic proof of an outer world will never be gained, nor yet syllogistic proof that a God exists or listens to prayer ... Not by subtlety of thought, but by specific sense, do we gain any acquaintance with the realities of things: and the Soul is the specific sense in which we come into contact with God. Let us not deal more slightingly with its testimony, than with that of the Touch or the Taste.[118]

Thus arguing for the existence of the object of the soul, Newman moved to the personal experience of God. Distinguishing between the masculine and feminine classes of the human being, he suggested that the higher spiritual states are reached not in the masculine but in the feminine mode, which may be available to men as well as women:

> That none can enter the kingdom of heaven without becoming a little Child, – guileless and simple-minded, is a sentiment long well known. But behind and after this there is a mystery, revealed to but few, which thou, oh Reader, must take to heart. Namely, if thy Soul is to go on into higher spiritual blessedness, it must become a *Woman*; yes, however manly thou be among men. It must learn to love being dependent; and must lean on God not solely from distress or alarm, but because it does not like independence or loneliness.[119]

Transgendering the soul as such (albeit relying on essentialist notions of masculinity and femininity), Newman described the relationship of the soul to God as one of feminine dependence, passivity and receptivity. Further, the relationship with God becomes not a matter of command but rather of desire and desirability as intimacy and not abstract law is the mode of operation between God and the soul. The soul becomes 'a partaker of the divine nature', melding with the divinity that she seeks.[120] The relationship of the soul and God is described as a 'marriage', and the love of the soul for God as like 'that borne by a woman to her husband'.[121]

In 'Hopes Concerning Future Life', Newman made clear that the purpose of religious devotion was the achievement of blessedness *in this life*. Hope in a future life might be justified by the experience of the divine but faith itself should not be based upon such a hope. In fact, Newman decried the ideological use of the promise (or threat) of an after-life as means for obtaining behavioral compliance from believers. In its place, Newman proffered the 'immanentization' of religious experience, its severance, for the most part, from otherworldly concerns or hopes. This was a theology, as Simon During puts it, 'for which God exists in this world not in any other'.[122] Anything else was impure, involving ulterior motives for belief.

The final chapter of *The Soul*, 'The Prospects of Christianity', is perhaps the most salient and instructive for a broader discussion of secularization and secularity. Here, as David Hempton points out, Newman anticipated our contemporary secularization debates.[123] However, I mean to argue more by this than I believe Hempton suggests. Newman sees traditional Christianity as dying within the urban centers of Europe. He presciently envisions that 'a real black infidelity will spread among the millions ... until the large towns of England become what Paris is'. Also, however, he argues for a '*new kind*' of religiosity, one that can appeal 'directly to the Conscience and the Soul'. Without such a religiosity, 'faith in Christianity once lost by the vulgar is lost for ever'. If Christianity depends upon 'arguments of erudition and criticism', then we must consider forever lost to it the 'tens of thousands who have learned to scorn Christian faith'. At present, 'Christianity has been turned into a Literature and therefore her teachers necessarily become a literary profession'.[124] But Christianity will not be saved by appeals to its factual, literary, or historical character. Instead, Newman recommended that to gain converts or to reconvert those atheists and pantheists who have fallen from the fold, modern-day apostles should:

> [i]n teaching about God and Christ, lay aside the wisdom of the wise: forswear History and all its apparatus: hold communion with the Father and the Son in the Spirit: from this communion learn all that is essential to the Gospel and still (if possible) retain every proposition which Paul believed and taught. Propose them to the faith of others, *to be tested by inward and spiritual evidence only*; and you will at least be in the true apostolic track.[125]

The historical apostles, Newman argued, never taught the inerrancy of any text or any set of doctrines. To follow the apostles meant teaching spiritual truth, not church or biblical inerrancy.

The reclamation of spirituality from the clutches of Christian doctrine and bibliolatry depended upon a division of labor between the mind and the soul, between the secular and the religious. As Hempton puts it: '[o]nly by rendering to the mind the things that belong to the mind and to the soul the things of the soul did Newman think that Christianity had any chance of surviving the onslaught of European secularization'.[126] To the mind would go all natural knowledge or 'external' evidences and to the soul would go direct spiritual experience or evidence. 'It is a first principle with us that the spiritual faculties discern spiritual things only, and cannot teach worldly and external truth, which essentially demands the aid of the specific bodily senses'.[127] On these terms, Newman believed that the spirit of religion might possibly still be saved in spite of the form that strangled it. However, '[h]e who knowingly sets Religion into contest with Science, is digging a pit for the souls of his fellow men'.[128] Thus, in 1849, Newman anticipated the notion of the non-overlapping magisteria (NOMA) of science and religion that Stephen Jay Gould formalized in the late twentieth century.[129]

Conclusion: *The Soul*, Secularism, secularity

Holyoake's interest in Francis Newman's religious writing, especially *The Soul*, was coterminous with the beginnings of the new movement and creed he called Secularism. In fact, in the issue of the *Reasoner* in which the term '*Secularist*' was first proposed by 'Edward Search' (William H. Ashurst) as a replacement for atheism and to refer to 'your phase of faith', and in which Holyoake also coined the term 'Secularism', Holyoake also published the second installment of his review of *The Soul*.[130] The interplay between Holyoake and Ashurst in this issue of the periodical suggests that discussions had already taken place between them regarding the new terms (Secular, Secularist, Secularism) as well the importance of Newman's writings for the formulation of the movement and creed. After first reading *The Soul* in 1849, in many issues of the *Reasoner*, Holyoake mentioned Newman's book or Francis Newman himself, or both. He contemplated the work for two years before finally writing a review.[131] In the interim and after Holyoake's review, correspondents to the magazine often referred to Newman and his work. It is clear from these many passages that the Holyoake branch of the freethought movement had adopted Newman's writing as part of its literature and strategy. In fact, we might say with no little justification that despite the criticism of Newman's theism in Holyoake's review, Secularism represented a fusion of artisan freethought and the new religiosity represented by *The Soul*. This is certainly how periodical opponents defending orthodox Christianity saw the matter. As the *Clerical Journal* put it in 1854:

> It [infidelity] now clothes itself in social respectability; it affects to have the welfare of the masses at heart, and proclaims a kind of religion of its own. From the pages of the elegant and amiable Leigh Hunt, to the more openly destructive volumes of Newman and Parker, we may see the same spirit at work.[132]

The uses made of Newman's work in the *Reasoner* certainly supported such assertions, as did Holyoake's many lectures on *The Soul*.

In a biographical and critical essay, *George Holyoake and Modern Atheism* (1855), the feminist freethinker Sophia Dobson Collet put the matter in more positive, yet emphatic terms. Characterizing the effect of Newman's *The Soul* on Holyoake and his subsequent development of Secularism, she wrote:

> Those Christians who are anxious for the conversion of Freethinkers would do well to study, in this little treatise, the remarkable effect which the faith of Francis Newman has produced on the mind of his Atheist reviewer ... And there can be little doubt that 'The Soul' has given a new tone to Mr. Holyoake's character. It has quickened into more distinct form all that was previously working in him towards noble development. In almost everything note-worthy which he has done or said since then, an attentive observer may trace the gracious influence of Francis Newman; – not in the form of servile imitation, but in that recasting of ideal aspiration, and that clearer perception and fuller development of high principles, which form the manliest tribute that one independent soul can pay to the excellence of another.[133]

Newman's *The Soul* allowed Holyoake to conceive of freethought in terms that superseded atheism and theism and included both atheists and theists or other religious believers. Newman's assertion that morality preexisted and was independent of religious creeds surely had much to do with this new, more capacious notion. No longer strictly concerned with negation, Secularism could conceive of a positive program inclusive of a moral system for the improvement of material conditions, taking hints from religionists themselves, who, like Newman, did not depend on religious doctrine for a moral system.

Newman's project also attracted George Henry Lewes and Thornton Hunt, the editors of the mid-century periodical *The Leader* (founded in 1850). One of *The Leader's* primary objectives was to promote a new kind of religiosity, one shorn of doctrine and yet appealing to the intuitive side of human beings so as to satisfy their spiritual natures. In his biography of Herbert Spencer, Mark Francis states that '[t]he author who most completely captured the spirit of this age was F. W. Newman, whose *Soul* and *Phases of Faith* were required reading for any self-conscious radical of the mid-century'.[134] Newman's works were central to George Henry Lewes's and Thornton Hunt's efforts at the *Leader* 'to find a foundation for its new religion', as part of a 'New Reformation'.[135] The *Leader* editors 'believed that "infidels" and "Atheists" were dated, and that they should aim their messages at the large and varied class of "spiritualists", who currently constituted a large proportion of the educated community'.[136] The datedness of atheism was not its only problem, however. They needed a new religion in order to capture the truth about human nature that religious feeling was inborn and that knowledge of this fact was necessary for the proper relationship to the universe. As the business manager of the *Leader* and regular contributor under the pen name 'Ion', Holyoake was well aware of this effort on the part of Lewes and Hunt, and their 'New Reformation' undoubtedly was significant for Secularism, which he was developing at the same time.

Not only did Newman's peculiar religiosity influence Holyoake but also Newman was deeply impressed with Secularism's philanthropic cast. Perhaps in emulation of Holyoake, Newman envisioned a Catholic Union of believers and unbelievers, which he conceived of in terms of a Church, whose only creed, if it could be so-called, would be philanthropy. This 'church of the future', Newman imagined, would embrace all denominations and sects of Christianity, as well as all the world's religions. It would explicitly include Secularists as well.[137] Like Holyoake, that is, Newman advanced an ideal for a union of religionists and non-religionists for the purposes of secular improvement; only Newman conceived of the construction from the religious side, while Holyoake conceived of it from the secular side.

Not only did Newman's conception of morality attract Holyoake, so did his spirituality. After his encounters with *The Soul*, Holyoake came to refer to the deity and the afterlife as outlying possibilities that might amount to bonuses to a Secular life lived well, rather than merely as beliefs held by antagonists. 'For if Secularism does not proceed upon the knowledge of a God *Actual*, it moves towards a God *Possible*'.[138] *The Soul* convinced Holyoake that an optative condition obtained and that belief in the deity remained a possibility even for a 'rational', intelligent, and educated person, although he himself would never declare any such belief. In fact, as we can see in Dobson Collet's biographical sketch, Holyoake would sit smack in the middle of the optative condition, affirming what might be affirmed, but no longer concerned with negating that which he might negate:

> Most of the original minds who commence active life on the side of Negation, come to alter their formal creed after some years' conflict with the realities of life. Many fail from want of personal or intellectual self-reliance, and turn to the Affirmations of Orthodoxy for moral support. Others, of less impressionable temperament, grow fanatic and impracticable, from the inability to perceive any truths beyond those which are peculiarly constitutional to themselves. But the healthiest and happiest of the band, escaping the Scylla of submission and the Charybdis of isolation, emerge safely into the Affirmations which are the true complements of their original Negations, and which, though long unanalyzed and but half perceived, have really been the central fountain from which that negative preaching derived all its generous life-blood.[139]

Here Dobson Collet noted the reconversions to orthodox Christianity of atheists and Secularists upon encounters with life's difficulties – or, that is, the reversibility of the secular that characterizes the optative condition of secularity. Holyoake, however, would never reverse his course as such. He remained an unbeliever, probably an agnostic, for life. However, he would take truth wherever he found it – in the writings and discussions of religionists like Francis Newman, and even the writings and sermons of Francis's brother, John Henry.[140] As such, his respect for believers had grown and deepened. He had discovered something that he considered more important than the differences that divided believers and unbelievers, belief and unbelief – a commonality, which he identified as the search for a single truth and morality:

> If we do but pierce beneath the antagonism from which all development issues, we shall see how, both with the Christian and the Freethinker, the same intention is ever at the bottom. We perceive a principle from different points, trace it to different roots, explain it

> in a different language, maintain it for different reasons, and foresee for it different conclusions: but the conflict continued, sometimes darkly, sometimes wildly, is for *one* morality and for *one* truth; and if there be in the end a Judge who looks with an equal eye on all, he will not fail to discern the motive and pardon the means.[141]

This was arguably the closest conciliation with religion in the history of freethought, and it was largely due to the impact of Francis W. Newman.

It was Holyoake himself who first wrote about 'The Three Newmans' when his Edwardian readers had only heard of the two. In a brief chapter in *Bygones Worth Remembering* (1905), Holyoake remarked that his primary focus, despite referencing the three brothers, would be on Francis: 'Though I name "three Newmans", this chapter relates chiefly to the one I best knew, Francis William, known as Professor Newman'.[142] I suspect, however, that Holyoake's emphasis was not merely due to his greater knowledge of Francis. Rather, his greater knowledge had depended on a prior, greater interest. To Holyoake, Francis Newman was simply the more compelling figure of the three. I share his fascination. Francis Newman became a central figure for Secularism for the same reason that I have presented him as a singular representative of secularity. He is a liminal figure, straddling a line between orthodoxy and unbelief, and refusing to opt for either of them. He thus represents the precariousness of the optative condition. He represents the 'religious', without the sacrifice of the 'rational'. Perhaps more than the committed atheist, he expressed the sentiments of his age and heralded those of the future: a commitment to spirituality unaccompanied by institutional loyalty and doctrinal orthodoxy. If the three Newmans together represent a triumvirate of secularity – the possibility of a pluralism of beliefs derived from a common religious background – of the three, Francis Newman's peculiar religious-cum-secular predicament characterizes secularity most signally.

6
George Eliot: The Secular Sublime, Post-Secularism, and 'Secularization'

Despite her personal skepticism and predominantly secular outlook,[1] we may regard George Eliot as a *post-secularist*. She was decidedly not a secularist of the Bradlaughian type. (See Chapter 3.) That is, she demonstrated a particular regard for religion and religious believers and generally acknowledged religion's ongoing viability, its potential to contribute to individual, cultural, and national identity and the general weal. Eliot often figured religion as a tissue that extended throughout and within the organic social body, a kind of living integument providing cohesion and shape, sustaining it in health and order. Religion could offer metanarratives that afforded meaning and coherence, ordering the experience of the subject, while enlarging the sympathies and recommending the dedication of individuals to broad social objectives. Eliot even acknowledged the Anglican Church as an important ecclesiastical body for its role in providing structural coherence and service to the community. And unlike other novelists of her time – such as Dickens and Trollope, who mercilessly caricatured clerical figures for hypocrisy, sectarianism, and factionalism – Eliot generally demonstrated respect for clerics and the clerical function, especially the pastoral duty of parish ministers. We have the 'saintly Mr. Tryan in *Scenes of Clerical Life*, "a powerful preacher, who was stirring the hearts of the people"; the eloquent and compassionate Methodist Dinah Morris in *Adam Bede*; the "wonderful preacher" Dr. Kenn in *Mill on the Floss*; the charismatic and increasingly self-deluded Savonarola in *Romola*; the learned and loquacious Dissenter Rufus Lyon in *Felix Holt*; and the affable *Farebrother* in Middlemarch'.[2] While indeed Eliot did subject Christian Britain and its ministers to criticism and unflattering comparison – for example, in her treatment of Mr. Gascoigne in *Daniel Deronda* – she recommended and endorsed the clerical ideal even in its failure. It repays us to recall that her first novel was entitled *Scenes of Clerical Life* (1857) and featured portraits of three Anglican clergymen.

Whether a pastor acted from a clearly defined Christian creed or a more loosely understood set of values for doing good was not of primary importance for Eliot. In Eliot's fiction, as Norman Vance has observed, '[m]oral and religious sympathy proves more durable than doctrine'.[3] Like her friend Francis Newman, whom Eliot referred to in a letter to Sara Hennell as 'our blessed St. Francis',[4] Eliot valued religious sentiment over theology, emotional truth over intellectual certitude, morality and generosity over 'correct' doctrines. As she wrote to Francoise D'Albert Durade in 1859, scarcely two years into her career as a novelist:

> I have not returned to dogmatic Christianity – to the acceptance of any set of doctrines as a creed, and a superhuman revelation of the Unseen – but I see in it the highest expression of the religious sentiment that has found its place in the history of mankind, and I have the profoundest interest in the inward life of sincere Christians in all ages. Many things that I should have argued against ten years ago, I now feel myself too ignorant and too limited in moral sensibility to speak of with confident disapprobation: on many points where I used to delight in expressing intellectual difference, I now delight in feeling an emotional agreement.[5]

Thus, we should not understand the representation of religion in George Eliot's novels as a matter of mere fictive construction or whimsical mediation on her part; rather, her novels illustrate a well-considered personal and historical understanding of the place of religion within a secular framework, and probably the period's finest expression of this understanding. This understanding is in fact what I mean calling her a post-secularist.

Eliot's sympathetic treatment of religion in fiction was essentially compatible with earlier and coterminous secular-religious projects for social and political amelioration, including Auguste Comte's 'Religion of Humanity', the *Leader*'s 'New Reformation' (see Chapter 5), and George Holyoake's Secularism.[6] As I have shown in Chapter 3, Eliot was part of the literary, intellectual, and scientific *avant garde* that helped George Holyoake to inaugurate and develop Secularism in the early 1850s, a group that included Thornton Hunt, George Henry Lewes, Harriet Martineau, Herbert Spencer, and others. She edited the *Westminster Review* and was part of the Chapman circle at 142 Strand, while also assisting George Henry Lewes with the *Leader*. These coteries were germinal to the founding and development of Secularism. Her close connections with her future partner, G. H. Lewes, as well as Hunt, Spencer, Martineau and others, means that, at the very least, she was cognizant of the formation of Secularism underway.[7] Indeed, we can figure Eliot as a Secularist of the Holyoake stripe. Welcoming religious believers to the tasks of secular improvement, in Comtean fashion, her novels imagined and suggested a kind of 'positive religion', not unlike that which Francis Newman described as the 'church of the future'.[8] (See Chapter 5.) Positive religionists cooperated with religious believers and might even contingently acknowledge their metaphysical convictions, although not necessarily taking the latter at face value.[9] Metaphysical assertions might either be ignored, or valued as ideals for promoting moral behavior and social cohesion. Such elements could function whether or not their epistemological status was accepted.

That is, while Eliot represented the persistence and appreciated the benefits of religion, she nevertheless repurposed the religious for generally secular ends, even if those ends remained in some sense transcendent. As Simon During puts it, 'Eliot's fictions mount an ambitious attempt at spiritual and intellectual invigoration and elevation, but one which does not adhere to revealed Christianity'.[10] Eliot's fiction represents a Victorian appropriation of the Romantic sublime registered in terms of social consequence rather than aestheticism. In Eliot's postsecular narratives, the salvation of the soul is transmuted into the salvation of character. 'Conversion' rehabilitates the subject for social duty as opposed to a heavenly destiny. The converted subject rejects personal egoism and narrow ambition and embraces her (generally minor) role as a contributing participant in the grand project of social amelioration and the slow, gradual development of a general human character.[11] The transcendental object of religious belief becomes the sublimity of secular causality stretching to eternity, the vision of which is only available in time. While the convert does not worship the secular sublime, she nevertheless is in awe of its magnitude and humbly submits to its power. In the final two paragraphs of *Middlemarch* (1872), Eliot conveys this sense of the secular sublime as encountered by its young heroine, Dorothea Brooke:

> Certainly those determining acts of her life were not ideally beautiful. They were the mixed result of young and noble impulse struggling amidst the conditions of an imperfect social state, in which great feelings will often take the aspect of error, and great faith the aspect of illusion. For there is no creature whose inward being is so

> strong that it is not greatly determined by what lies outside it. A new Theresa will hardly have the opportunity of reforming a conventual life, any more than a new Antigone will spend her heroic piety in daring all for the sake of a brother's burial: the medium in which their ardent deeds took shape is forever gone. But we insignificant people with our daily words and acts are preparing the lives of many Dorotheas, some of which may present a far sadder sacrifice than that of the Dorothea whose story we know.
>
> Her finely touched spirit had still its fine issues, though they were not widely visible. Her full nature, like that river of which Cyrus broke the strength, spent itself in channels which had no great name on the earth. But the effect of her being on those around her was incalculably diffusive: for the growing good of the world is partly dependent on unhistoric acts; and that things are not so ill with you and me as they might have been, is half owing to the number who lived faithfully a hidden life, and rest in unvisited tombs.[12]

This is where Eliot had left religiosity and the prospect of human agency by the end of her penultimate novel. No new Saint Theresa could again emerge under the conditions of the last quarter of the nineteenth century, by the time philological and philosophical secularism had done so much of their cultural work,[13] a period during which scientific naturalism had risen to prominence if not exclusive domination, and by which time the cohesiveness of a religious worldview had been severely fractured. 'Great faith' (in anything) took 'the aspect of illusion'. Likewise, the prospects for 'the ardently willing soul'[14] had been profoundly altered. Spirituality had to be recalibrated for tasks with secular aims and relative insignificance, since the stakes were no longer the glory of God or the rewards of eternal salvation, but rather the arduous making of a better world. For such intensely passionate and devotional souls as Dorothea Brooke, who, as Eliot leads us to believe, likely would have become a nun in a former age, the adjustment was significant.[15]

Thus, the post-secular is not a return to religion from secularism. Rather, it describes a condition of secularity under which the modality of religiosity has been altered by the secular and relativized as one possibility among others, a relativism that profoundly affects and 'fragilizes' it.[16] Post-secularism signifies the persistence of religion, but religion that has become a choice among other options. The very structure of belief has changed under the post-secular condition. Religious belief, when it survives, is contingent and unstable, and in some cases, cannot help but become 'secular', as was the case for Dorothea and George Eliot.

Yet, in *Daniel Deronda*, her final novel, George Eliot gestured toward an even broader and deeper sense of post-secularism or secularity than suggested above. We might say that in this career-crowning novel, Eliot took a surprising, while somewhat ambiguous 'religious turn'. While retaining secular functions for religion, *Daniel Deronda* reserves a greater role for religion than Eliot had accorded it in previous novels, and the religion that it does represent is decidedly more resilient and central to the novel and its vision. In *Daniel Deronda,* religion functions at levels both *above* and *below* the secular. By 'below' the secular, I mean that, as the source of cultural, ethnic, social, and textual identity and cohesion,[17] Judaism represents a resistance to the forces that threatened to undermine the character of the British nation, leaving it awash in a rootless cosmopolitanism and homogenous secularization, as seen in the 'English half' of the novel.[18] Judaic religiosity, Eliot suggests, resists the secular by occupying precisely the space that the secular would otherwise have taken up. The 'Jewish' plot line compresses the quotidian and the historic, the ordinary and mythic, the 'flesh' and the 'beloved ideas' of spirituality.[19] By 'above' the secular, I mean that the novel points to the possibility of transcendence. It folds the discrete, particular, and accidental events and characters of its 'Jewish half' into the cohesive, universal, messianic promise of Judaism, a construction that betrays such improbability that it begs belief in an extraordinary agency, at least on the part of some of its characters, and perhaps its readers too. This elevation of plot requires a new kind of generic accommodation; a post-realist or realist-romance narrative must register the improbable as credible by virtue of the religious authorization of events. Placed against the realist plot of the 'English half' of the novel, the religious narrative requires that the reader also suspend disbelief. This difficulty is mitigated by the traveling between the two parts by the itinerant, eponymous hero, Daniel Deronda, whose interventions on behalf of Gwendolen Harleth – the beautiful, willful, and self-consumed hero of the 'English half' – stand as a metonym for the possibilities of spiritual regeneration in general.

In *Daniel Deronda,* Eliot suggests that Judaism, by resisting the onslaughts of rootless cosmopolitanism and secularization, serves as a potential paragon for the reinvigoration of the spiritual life of the British nation.[20] This construction reverses the stereotypical notion of 'the Jew' as the rootless cosmopolitan whose presence represents a threat to the supposed racially homogenous national populace. *Daniel Deronda* also critiques and subverts the Christian appropriation of Jewish religious identity, the *religiophagism* of Christianity. It effectively reverses the relationship that traditionally obtains between the two world religions, wherein the Christian understands itself as the ultimate embodiment and transcendence of the Judaic.[21] In *Daniel Deronda*, the Judaic both subtends and transcends the Christian as Judaism is restored to its historical place as the fount of Christianity, while promising to serve as a model for the latter. Eliot re-appropriates what is perhaps the most profound mystery in Christianity, the incarnation of God, as the Jewish people represent the literal embodiment of spiritual knowledge in a historic, now dispersed, and future nation. If, as Graeme Smith has suggested, secularization has really amounted to Christianity masquerading as secularism, then in *Daniel Deronda*, Judaism survives both.[22]

Eliot's *Daniel Deronda* deftly characterizes post-secularism or secularity in the third quarter of the nineteenth century – a space where the religious and the secular subsist as coexisting and complementary others, yet one in which religion in the form of Judaism represents a metonymic challenge to the prospects of secularization. Judaism, I argue, serves in the novel as an exemplar of religious persistence, and one that, as Eliot sees it, Christianity can emulate.

I begin in this chapter by exploring the notion in *Daniel Deronda* of religion as a persistent cultural, ethnic, literary, and social identification that inheres within and between Jewish subjects, and which allows them and Judaism itself to resist not only appropriation by religious otherness but also by the secular. I then treat the transcendent character of religion as portrayed in the novel, and continue by examining the question of Judaism and 'secularization' in late nineteenth-century Britain.

Religion and 'blood'

Daniel Deronda comprehends religion as other than belief per se, as a formation that is more than the mere antithesis of reason. In *Daniel Deronda*, religion is not merely 'a set of propositions' but rather resides in the very 'fibre' of individual and social being: 'Mirah's religion was of one fibre with her affections, and had never presented itself to her as a set of propositions'.[23] This sense of religious belief is what the contemporary political theorist William E. Connolly calls the 'visceral register of subjectivity and intersubjectivity'. Both secular and religious commitments, Connolly argues, should be understood in terms of a sensitive complex of affect, thought, and judgment that is at once pre-cognitive and cognitive, and which is shaped by a thick 'host of historically contingent routines, traumas, joys and conversion experiences [that] leave imprints upon the visceral register of thinking and judgment'.[24] Both the religious and the secular are understood as lying beneath the level of rationality and yet as partaking in rationality as well. George Eliot very much comprehended religion as operating at this subjective-intersubjective register, and in *Daniel Deronda*, she ascribed this sense of religiosity to Judaism.

In *Daniel Deronda*, Eliot's trope for what Connolly calls the visceral register is 'blood'. In general, 'blood' is a complex figure representing connection, communication, health, inheritance, lineage, nation, passion, and race. In *Daniel Deronda*, 'blood' simultaneously gestures toward the physical body and spirituality; it is a nebulous marker that slips between quasi-racial and cultural categories. The identity of the Jewish people is a paradox for the novel, defined variously in terms of culture, history, literature, race, and religion, the constellation of which is concretized in the figure of 'blood'. An ambiguous yet remarkably predictable signifier, 'blood' functions as a reliable index for reading the moral economy of *Daniel Deronda*, while also representing the inversion of religious and social hierarchies.

Daniel Deronda, as I have suggested, consists of two related but nevertheless distinct and intertwining narrative threads. One tracks the beautiful, defiant, and conceited Gwendolen Harleth, and the other the honorable, introspective, and soulful Daniel Deronda. Deronda is the ward of the baronet, Sir Hugo Mallinger, whom he calls his uncle but secretly believes to be his father. Unsure of his parentage, yet raised as a gentleman, Deronda attended Cambridge for a time, and lives with Sir Hugo and Lady Mallinger at Sir Hugo's estate, the site of the ruins of a medieval abbey. (The overwriting of the Abbey as Sir Hugo's estate signifies the state of Christianity within British secularity; the old chapel has become a horse stable.) Graced with a rich interior life, poetic sensibilities, and broad sympathy, Deronda nevertheless lacks a vocation that would 'compress his wandering energy'[25] and provide a purpose appropriate to his elevated moral and affective disposition. Gwendolen, along with her mother and four 'half-sisters', has recently moved to Offendene, a house chosen for its proximity to Gwendolen's uncle, the Anglican clergyman, Mr. Gascoigne. Gwendolen reigns supreme among her sisters and mother in a 'domestic empire',[26] which has recently expanded to include her two cousins Rex and Anna, the children of Mr. and Mrs. Gascoigne. Dreading the prospect of marriage, however – which signals a potential forfeiture of the extended sovereignty and willful dominance that she has come to expect – Gwendolen has remained aloof and unattainable to suitors. Yet, marriage appears necessary to attain the wealth and triumph that she imagines as her birthright. Gwendolen does not fare well. Her marriage to the rich, imperious Mallinger Grandcourt quickly becomes a gilded cage with Grandcourt as her jailor – until, that is, she is released by Grandcourt's sudden death in a sailing accident in the Mediterranean, a drowning that Gwendolen witnesses but fails to prevent. The unfolding of Daniel's pilgrimage, on the other hand, leads to both personal fulfillment and a higher social and religious calling. After learning of his Jewish parentage, he embraces a Jewish identity and Ezra Mordecai Cohen's vision of a New Judea. In the end, he marries Mirah, Mordecai's sister, and removes to Palestine to begin efforts to establish a Jewish nation. Gwendolen secretly entertains the possibility of a love relationship with Deronda until the penultimate chapter, when Deronda informs her of his newfound Jewish identity, his Judaic project, and his plan to marry Mirah. Gwendolen's devastation is profound. She is thrust into an encounter with the overwhelming magnanimity of 'the wide-stretching purposes' of the world – a secular sublime – 'in which she felt herself reduced to a mere speck'.[27] Daniel, on the other hand, has found his 'ideal task', a 'social captainship', in which he felt himself 'the heart and brain of a multitude'.[28]

When Daniel first sees Gwendolen, she has not yet married Grandcourt, and Deronda has not learned of his high calling. In the fictional city of Leubronn, Deronda first lays eyes on Gwendolen as she sits at a gaming table. On a winning spree, until Daniel's gaze seems to reverse her fortune, her defiant nature is on full display.[29] We later learn that she has recently absconded from Offendene, after a harrowing encounter with the mysterious and foreboding Lydia Glasher, Grandcourt's former lover and the mother of his children. Lady Glasher has warned Gwendolen not to marry Grandcourt, as doing so would represent a crime against herself and her children, and a life rightfully attended by guilt. Yet in this first chapter, the reader knows nothing of this, only that Gwendolen and Deronda are complete strangers and that Gwendolen suddenly finds herself caught in the act under Deronda's observation. Her spiritual condition is measured in blood, or the lack thereof: 'It [her sense of his judgment] did not bring the blood to her cheeks, but sent it away from her lips. She controlled herself by the help of an inward defiance, and without other sign of emotion than this lip-paleness turned to her play'.[30] Embarrassed by a sudden awareness of inferiority under Deronda's smarting, supervisory gaze, her pale lips betray an awareness of her relative moral and spiritual disadvantage, which she hopes to mask under her usual cloak of superciliousness.

Blood connects the novel's characters to familial lineage, as in the inheritance plot, according to which the morbidly disaffected, perennially bored Grandcourt is the likely prospective heir to Sir Hugo Mallinger's vast legacy, including his title, while Deronda has lived as the latter's ward, thinking for years that he is his son. This bloodline represents the spiritual and moral degeneration of Britain and the depth to which its ruling class has sunk. Relying solely on its material power, it has extended its empire, but, as in the person of Grandcourt, it has become 'pale-blooded'[31] and soulless.[32]

Deronda's identity is also dependent on bloodline. His discovery of his Jewish identity in a meeting with his mother allows his choice of a Jewish identity, his embrace of Judaism, his marriage to Mirah, and his project of working toward the establishment of a Jewish state in Palestine. I will discuss the meaning of 'blood' in this connection, below.

'Blood' also potentially links the novel's characters to a community and a local history, allowing them to resist a universalizing cosmopolitanism, or the secularizing, homogenizing forces of modernity that threaten cultural identity and ethical values, sometimes before such values can even be imparted. A conspicuous invisibility or absence of blood, or its lack of nurturance, signals a failure of socialization and results in the deformation of character. In a passage that breaks the narrative frame, the narrator opines that Gwendolen's moral and spiritual vacuity is due to a lack of rootedness in place, a rootedness that would have provided her ethical standards and a cultural identity like a second nature inhering in the 'blood':

> Pity that Offendene was not the home of Miss Harleth's childhood, or endeared to her by family memories! A human life, I think, should be well rooted in some spot of a native land, where it may get the love of tender kinship for the face of earth, for the labours men go forth to, for the sounds and accents that haunt it, for whatever will give that early home a familiar unmistakable difference amidst the future widening of knowledge: a spot where the definiteness of early memories may be inwrought with affection, and kindly acquaintance with all neighbours, even to the dogs and donkeys, may spread not by sentimental effort and reflection, but as a sweet habit of the blood.[33]

This passage recalls Eliot's previous novel, *Middlemarch*, in which the heroine, Dorothea Brooke, a self-sacrificing and morally defensible character for Eliot, does feel connected to the community and the pulse of life, which extends like the landscape to the horizon of her future:

> On the road there was a man with a bundle on his back and a woman carrying her baby; in the field she could see figures moving – perhaps the shepherd with his dog. Far off in the bending sky was the pearly light; and she felt the largeness of the world and the manifold wakings of men to labour and endurance. She was part of that involuntary, palpitating life, and could neither look out on it from her luxurious shelter as a mere spectator, nor hide her eyes in selfish complaining.[34]

Both passages represent an ideal found in Wordsworth – a childhood rooted in (usually rural) places yields strong affections that serve as the basis for adult character:

> These beauteous forms,
> Through a long absence, have not been to me
> As is a landscape to a blind man's eye:
> But oft, in lonely rooms, and 'mid the din
> Of towns and cities, I have owed to them,
> In hours of weariness, sensations sweet,
> *Felt in the blood,* and felt along the heart;
> And passing even into my purer mind
> With tranquil restoration: – feelings too
> Of unremembered pleasure: such, perhaps,
> As have no slight or trivial influence
> On that best portion of a good man's life,
> His little, nameless, unremembered, acts
> Of kindness and of love.[35]

Gwendolen's childhood lacked this very 'influence' – '[b]ut this blessed persistence in which affection can take root had been wanting in Gwendolen's life'[36] – thus, she is self-indulgent and narcissistic; her interest in others is limited to their ability to please her. 'My plan is to do what pleases me', she peremptorily announces to Rex.[37] Although she does not know it, Gwendolen's moral system is Benthamite Utilitarianism; her guiding rule is, as Lisa Bonaparte has pointed out, the 'hedonic calculus' of pleasure.[38] Eliot aims to show that this moral system leads not to happiness for the greatest number but rather to misery for the individual. Like the 'New Reformation' critics of Benthamite Utilitarianism, Eliot critiqued utilitarianism for its lack of regard for duty and virtue.[39] For Eliot, the moral sense was intuitive rather than rational. Eliot tests the pleasure principle in fiction, immersing her *de facto* utilitarian in the consequences of a faulty moral system. Gwendolen's *ennui*, the narrator suggests, results from a moral incapacity or lack of training in empathy, an absence of the moral sense and a higher motive that would relieve her of the bondage of self. The impulse to pleasure remits pain. Rather than cherishing the familiar, she treats it with derision, which leaves her in a state of perpetual dissatisfaction. This is evidenced by her handling of Rex, the lovelorn cousin whose advances she rejects and whose feelings she tramples upon. Lacking any positive regard for the value of others (with the possible exception of her mother), incapable of being pleased by suitors, she declares her petulant and hazardous disenchantment, boldly announcing to her mother: 'I shall never love anybody. I can't love people. I hate them'.[40]

A similar although hardly identical rootlessness in Deronda signals an impartiality that precludes moral conviction and determined action, before, that is, he discovers his Jewish bloodline and identity. However, unlike Gwendolen, in Deronda's case, his impartiality results from too much, not too little sympathy:

> His early-wakened sensibility and reflectiveness had developed into a many-sided sympathy, which threatened to hinder any persistent course of action: as soon as he took up any antagonism, though only in thought, he seemed to himself like the Sabine warriors in the memorable story – with nothing to meet his spear but flesh of his flesh, and objects that he loved. His imagination had so wrought itself to the habit of seeing things as they probably appeared to others, that a strong partisanship, unless it were against an immediate oppression, had become an insincerity for him. His plenteous, flexible sympathy had ended by falling into one current with that reflective analysis which tends to neutralise sympathy ... A too reflective and diffusive sympathy was in danger of paralyzing in him that indignation against wrong and that selectness of fellowship which are the conditions of moral force; and in the last few years of confirmed manhood he had become so keenly aware of this that what he most longed for was either some external event, or some inward light, that would urge him into a definite line of action, and compress his wandering energy ... But how and whence was the needed event to come? – the influence that would justify partiality, and make him what he longed to be yet was unable to make himself – an

> organic part of social life, instead of roaming in it like a yearning disembodied spirit, stirred with a vague social passion, but without fixed local habitation to render fellowship real?[41]

Deronda's lack of partiality, or his impartiality, while owing partly to his particular character, is a condition of the nation, and one that has also already been blamed for Gwendolen's insolence.[42] The narrator reveals that Deronda partly faults his upbringing, 'which had laid no special demands on him and given him no fixed relationship except one of a doubtful kind'.[43] This lack of attachment represents the condition of a cosmopolitan, secularized nation. But the passage also seems to describe the 'cosmopolitan indifference' that Eliot's persona in 'The Modern Hep! Hep! Hep!' – the final essay of Eliot's final book – attributes to the 'expatriated, denationalized race' of Jews, who would be tempted to 'drop that separateness that is made their reproach'.[44] That is, the same rootless cosmopolitanism that had been historically associated with 'the Jews' is a condition that can obtain for any group or nationality – English, Jewish, or Italian, for example, and that does obtain to a degree. Given his gentlemanly English upbringing, Deronda is a rootless cosmopolitan. But in his case, the condition also stems from lack of knowledge of and connection to familial roots. Before he learns of his heritage, he is effectively a member of the 'expatriated, denationalized race' of Jews, owing to his mother's decision to have him raised by an English gentleman. In the 'Modern Hep! Hep! Hep!', Eliot's persona suggests that the eventuality of such cosmopolitanism is inevitable in the long term. 'The tendency of things is toward the quicker or slower fusion of races. It is impossible to arrest this tendency', but this tendency must be guided and moderated so that a national culture can do its work.[45] In the present historical moment, the 'blood' of national culture is required to produce character and morality, and this is no less true for Deronda. Of course, Deronda's eventual discovery leads him to his national culture, which gives him the direction required for determinate, moral action.

Mirah Lapidoth, Mordecai's sister, represents a striking contrast to both Gwendolen and the early Deronda. Mirah is a distressed and beautiful 'Jewess' whom Deronda saves from a potential suicidal drowning in the Thames and places under the care of Mrs. Meyrick and her three daughters, the mother and sisters of his Cambridge friend, Hans. Mirah is the novel's paragon of virtue. Despite being a wandering Jew (like Deronda will become), Mirah's cultural rootedness is synonymous with a moral resolution and firm religious and personal identity. If Gwendolen is the 'double and satirical', cynical and 'spoiled child',[46] Mirah has far greater cause for cynicism. However, she is not cynical. 'For Mirah was not childlike from ignorance: her experience of evil and trouble was deeper and stranger than his [Deronda's] own'.[47] Despite the fact that her father (Lapidoth) had effectively kidnapped her in Prague, separated her from her mother, transported her to America, shuttled her onto the stage to earn money, and planned to sell her into concubinage, she retains a religious conviction stemming from entrenched filial affection. For Mirah, keeping the Judaic faith is tantamount to keeping faith with her loved and lost mother. While Gwendolen 'always disliked whatever was presented to her under the name of religion, in the same way that some people dislike arithmetic and accounts',[48] for Mirah, the practices of her religion are associated with 'the feelings I would not part with for anything else in the world'.[49] Mirah deflects charges from Amy Meyrick that in the synagogue, women are relegated to a subordinated position: 'Excuse me, Mirah, but *does* it seem quite right to you that the women should sit behind rails in a gallery apart?' Mirah, not understanding the implication, responds with surprise: 'Yes, I never thought of anything else'.[50] Yet, Mirah's religious conviction is not the result of a protracted indoctrination, her abduction having mostly curtailed her religious education. '"She says herself she is a very bad Jewess, and does not half know her people's religion," said Amy, when Mirah was gone to bed'. Amy continues, echoing the rhetoric of conversion societies in Britain, such as the Society for the Conversion of the Jews,[51] in expressing a hope that Mirah's Judaism 'would gradually melt away from her, and she would pass into Christianity like the rest of the world, if she got to love us very much, and never found her mother. It is so strange to be of the Jews' religion now'.[52]

But abandoning her religion is inconceivable to Mirah. Mrs. Meyrick discovers this when she suggests that 'if Jews and Jewesses went on changing their religion, and making no difference between themselves and Christians, there would come a time when there would be no Jews to be seen'. Mirah objects passionately: '"Oh please not to say that," said Mirah, the tears gathering. "It is the first unkind thing you ever said. I will not begin that. I will never separate myself from my mother's people."'[53] Her religion is in her 'blood'.

The most complex relationship to 'blood' in connection with religion involves Deronda. The question of biological determinism is central to the novel's treatment of Deronda's eventual embrace of Judaism and his acceptance of the Judaic mission bequeathed by Mordecai. A central question is whether Deronda's attraction to Judaism and his acceptance of Mordecai's mission for the establishment of Israel as a nation is racially determined; is his religiosity the function of a quality inhering in the 'blood' – racially construed? It is clear that the Jewish plot depends on whether or not Deronda has 'Jewish blood',[54] but it does not necessarily follow that the presence of Jewish blood guarantees his acquiescence, that is, that the novel follows a racially determined plot line in the 'Jewish half'. The question has important implications regarding just what the novel makes of religion within modernity. Does the 'racial' character of Judaism represent an exception to what the narrator of *Daniel Deronda* figures as the otherwise largely secularized modern world; or, is Judaism, rather than representing an exception to modern religion, offered as an exemplar for religiosity generally? If Christianity might emulate Judaism in some sense, then, according to the novel, Judaism's supposed racial basis is not the differentiating factor for the survival of religion generally, as religion is not dependent upon race. Such survivability would have significant implications regarding the novel's and Eliot's position on what we now understand as secularization. Is the inevitable 'fusion of races' equivalent to secularization? The answers lie in the novel's representation of the connection, whether extremely tight or relatively loose, between race and religion. Just what does Eliot mean by the figure of 'blood'?

In his biography of George Eliot, Frederick Karl suggested that in 1873, Eliot was wrestling with such questions before writing *Daniel Deronda*:

> But as she approached her final long fiction, she was still attempting to find some middle path through all the minefields of nineteenth-century beliefs: Comtean and Harisonian positivism, religious orthodoxy of one kind or another, utilitarianism (Bentham's or John Stuart Mill's), Huxley's agnosticism, Darwin's evolutionism and determinism, Spencer's social Darwinism, and Marxism and its various offshoots.[55]

In *Daniel Deronda*, Eliot deals with such belief systems in the Hand and Banner, the pub where 'The Philosophers' club meets. This meeting of working-class intellectuals, 'who had probably snatched knowledge as most of us snatch indulgences, making the utmost of scant opportunity',[56] allows Eliot to probe current philosophical perspectives – including Godwinian social environmentalism, social Darwinism, and 'rational' Judaism – and to dispense with them handily.[57] The topic, 'the law of progress', allows Lilly, the Jewish copying-clerk, to argue that progress is actually 'development' (evolution), while Miller, the Germanic second-hand bookseller, claims that ideas are the ruling forces in the world, and the Englishman and wood-inlayer, Goodwin (recalling Godwin), counters Miller by retorting that successful ideas are actually practices embedded in materiality.

Three of the Jewish characters introduced in the Hand and Banner, Lilly, Pash, and Gideon – the last an optical instrument maker and proponent of 'making our [Jewish] expectations rational'[58] – may very well have been based on the person of Alfred Gutteres Henriques. Henriques was a prominent Anglo-Jewish barrister, Vice-President of the Anglo-Jewish Association, deputy lieutenant for the City of London,[59] and author of a legal text on land credit and mortgages.[60] He was also a pro-Darwinian evolutionist and member of the Reformed West London Synagogue[61] (founded in 1870). In 1869, he argued in a letter to the editor of the *Jewish Chronicle* that the 'Law of Evolution' thoroughly outmoded traditional Judaism and that in response to scientific rationality the latter needed to be radically reformed and made 'rational'.[62] As Geoffrey Cantor has observed, 'Henriques was seen to pose a significant threat to the community and to Jewish tradition'.[63] A well-known figure in Jewish circles, the prominent and successfully assimilated Henriques may have served as a source for Eliot, although views such as his were placed in the mouths of a plebeian cast. Yet another possible source may have been Raphael Meldola, the Jewish professor of chemistry at the Royal College of Chemistry and later at Finsbury Technical College. Meldola was a friend and correspondent of Charles Darwin, an outspoken Darwinist, a Fellow of the Royal Society, and 'the most eminent Jewish naturalist of the period'.[64] Meldola authored a number of papers explaining insect morphology and adaptation and other issues deploying the Darwinian principle of natural selection, and contributed articles to *Nature* where he forcefully expressed his Darwinian commitments. He also translated August Weismann's *Studien zur Descendenz-Theorie* (1875) into English as *Studies in the Theory of Descent* (1882). The translation included a preface by Darwin. Interestingly, Meldola also lectured Jewish workingmen in science, and his father was an optician, likely served by an optical instrument maker like Gideon. But Meldola was also a committed Jew, although not a very observant one.[65] Of course, Eliot may have drawn from a number of such sources as Henriques and Meldola for these pro-Darwinian and determinist characters.

Targeting the application of evolution to the social (social Darwinism), the ultimate determinism, Deronda directly contradicts Lilly, stating: 'there will still remain the danger of mistaking a tendency which should be resisted for an inevitable law that we must adjust ourselves to, – which seems to be as bad a superstition or false god as any that has been set up without the ceremonies of philosophizing'. Mordecai agrees: 'That is a truth ... Woe to the men who see no place for *resistance* in this generation!'[66] Thus, the two speakers, whom the novel invests with authority – especially in consideration of the observation that their opposing interlocutors 'had probably snatched knowledge as most of us snatch indulgences, making the utmost of scant opportunity' – flatly reject determinism, determinism of the kind presented: the social world understood as the result of an inevitable law of development or causality of which social actors are the necessary products. As Lisa Bonaparte points out, 'Eliot ... cannot agree that the course of human events does not depend on human beings'. Instead of such materialist determinism, the novel endorses idealism, in both senses of the word – the priority of ideas over matter (philosophical idealism) and the possibility of human perfectibility (political idealism).[67] Further, the fact that three of the six philosophers in the club are Jewish but do not accept Mordecai's vision shows that 'race' does not determine political or religious outlook. Even Mordecai's supposed racial determinism is qualified by this fact. While Mordecai may *believe* in racial determinism ('because I was a Jew ... because I was a Jew'[68]), where Judaic belief is concerned, the novel undermines this position.

Probably the most important narrative element bearing on the question of determinism in the novel is the series of two meetings in Genoa between Deronda and his mother, the Princess Leonora Halm-Eberstein. In this dramatic sequence, Deronda comes to know his mother for the first time and learns of his Jewish ancestry. Deronda eagerly adopts the Jewish identity that his mother had rejected; this leads some scholars to argue that Deronda's acceptance of his heritage is represented as inevitable. For example, in his *Literary Secularism*, Amardeep Singh suggests that Deronda's embrace of his Jewishness and the Judaic faith is 'decidedly *not* voluntary'.[69] The implication is that Deronda's bloodline produces an inevitable adherence to Jewishness and Judaism, and ultimately, to Mordecai's Jewish nationalism. Others have analyzed the Leonora narrative in terms of Leonora's resistance to Jewish patriarchy. This line of thinking is suggestive for my discussion. Susan Meyer has argued that like the return of the repressed, 'Leonora's escape [from the life of a Jewish woman as dictated to her by her father] expresses the impulses the novel is trying to suppress'.[70] That is, the Leonora narrative is the novel's otherwise suppressed expression of the possibility for women to escape the patriarchal determination of their identities and fates. My point here is not to treat Leonora's narrative as Eliot's 'suffocated and shrivelled'[71] vision of women's self-determination, although the point is well taken. As I see it, Leonora's story more prominently works to suggest, while simultaneously subverting, the notion of the racial determination of Jewish identity and religiosity. Although Deronda is utterly *compelled* to adopt his new identity, Leonora's historical refusal of the same – which has resulted in Deronda's personal history to this point – makes clear that Daniel's accession is indeed voluntary. Leonora does feel the weight of her father's Jewish legacy and his will that his grandson carry it forward. Thus, she has decided to beckon Deronda and inform him of his lineage. But her own life, her choice to become an artist and cosmopolitan (like Klesmer), rather than a Jewish woman and mother, demonstrates what Eliot holds forth as an essential condition of any meaningful morality and religiosity: the possibility of resistance. Whatever else 'blood' signifies in the novel, it does not mean the determination of religiosity by race. 'Blood' is a necessary condition for Deronda's election, but it is not a sufficient one.

As Virgil Martin Nemoianu argues, *Daniel Deronda* bears evidence of Eliot's extensive engagement with Baruch Spinoza's ethics and conception of human freedom. Eliot's intimate understanding of Spinoza was owing to her work translating his *Tracticus Theologico-Politicus* (1670) in 1848, as well as his *Ethics* (1677) in 1854.[72] Freedom for Spinoza did not mean the willy-nilly assertion of self-will, the imposition of one's desires onto the world, as if the individual were separate from the social body upon which she acted. Under such a conception, 'one's actions would have no necessary connection to the world; they would be arbitrary'.[73] This is the notion of freedom that Gwendolen has professed and acted upon, and one that the novel is at pains to declare mistaken. For Spinoza, on the other hand, freedom depends on a kind of determinism.

How are these two elements – freedom and determinism – compatible? Spinoza's conception of freedom involves the individual human agent in a quest for knowledge of self-in-world, knowledge of the multiple social determinations acting upon the self in a thick network of causal relations, and the election of 'rational' actions based on such knowledge. 'For a human being to be free, the desires which determine her behavior must arise from rational knowledge of herself and her situation relative to other human beings and the rest of what is. Only then can she act coherently and effectively, in a way that benefits her'.[74] Deronda's adoption of Jewish identity (and the train of convictions that follow from it) is conditioned by the multiple, existing social determinations that have acted upon him, and which inform his actions with reference to them. His agency is constrained by the concrete particulars that comprise his embeddedness within the social whole, but his 'freedom' consists of an informed response in relation to them. That is, he has acted freely in the Spinozan sense.

While the novel supports this view throughout, a moment that is crucial to its view of freedom is the discussion between Mordecai and Deronda after the latter has revealed his Jewish identity to Mordecai and Mirah. Learning of Deronda's ancestry, and knowing that he will soon die, Mordecai wastes no time in vehemently urging 'the marriage of our souls', the 'transmission' between himself and Deronda that he has spoken of before. As part of the melding of their identities, Mordecai suggests that Deronda become the author of his writings: 'For I have judged what I have written, and I desire the body that I gave my thought to pass away as this fleshly body will pass; but let the thought be born again from our fuller soul which shall be called yours'.[75] The first 'body' Mordecai refers to here is the body of work that he has authored. This body, he suggests, must be 'born again' in Deronda, just as Mordecai's earthly body is passing away. That is, Deronda must body them forth again, having engulfed and been engulfed by Mordecai's identity.[76]

While Deronda has been zealously enthusiastic in this encounter to this point, he now senses an infringement and demurs. Mordecai's declared intention to foist his identity on Deronda (in the form of Mordecai's oeuvre which Deronda must body forth) precludes Deronda's exercise of self-determination and entails his usage in a ventriloquism that he cannot agree to in advance:

> 'You must not ask me to promise that', said Deronda, smiling. 'I must be convinced first of special reasons for it in the writings themselves. And I am too backward a pupil yet. That blent transmission must go on without any choice of ours; *but what we can't hinder must not make our rule for what we ought to choose*. I think our duty is faithful tradition where we can attain it. And so you would insist for any one but yourself. Don't ask me to deny my spiritual parentage, when I am finding the clue of my life in the recognition of my natural parentage'.[77]

This is perhaps the most complex passage in the novel. It represents the Spinozan conception of freedom, only reworked significantly by Eliot and made more paradoxical, subtle, and profound than it was in Spinoza's hands. It defines freedom not in terms of rational action but rather the possibility of refusal even in the face of inevitability.

Although Deronda concedes that the 'blent transmission must go on without any choice' of his own or Mordecai's, he nevertheless reserves the right not to *choose* it – to refuse having his choice determined by necessity. Although he may not be able to hinder the eventuality, he maintains the possibility of withholding his consent to it. That is, he claims the prerogative of resistance that Mordecai had insisted upon in 'The Philosophers' meeting. The possibility of resistance, even in the face of inevitability, amounts to freedom. Deronda's reassertion of agency here suggests that even his very Jewish identification has been a matter of choice. Although he may be a Jew by 'blood', the significance of the fact falls to his discretion and depends on his interpretation. He makes this clear in his conversation with his mother, after she asks him whether he will 'turn [himself] into a Jew like him [his grandfather]?' Deronda responds:

> 'That is impossible. The effect of my education can never be done away with. The Christian sympathies in which my mind was reared can never die out of me', said Deronda, with increasing tenacity of tone. 'But I consider it my duty – *it is the impulse of my feeling – to identify myself, as far as possible, with my hereditary people*, and if I can see any work to be done for them that I can give my soul and hand to, I shall *choose* to do it'.[78]

Deronda's identification is a choice that is his to make.

There are earlier indications that Mordecai has asked something of Deronda that he would not accept for himself, and that he regards his own religiosity not as an unwilled imposition over which he has no control, but rather as a matter of self-determination. I point again to the meeting of 'The Philosophers' club, when Mordecai erupts into an oration about the Jewish people and their future: 'I say that the strongest principle of growth lies in human choice. The sons of Judah have to choose that God may again choose them. The Messianic time is the time when Israel shall will the planting of the national ensign'.[79] Even the messianic promise of Judaism depends on the active, decided election of their historical and religious role by the Jewish people.

Therefore, in *Daniel Deronda*, 'blood' does not signify racial determination of religious identification and practice, or the necessity of Zionism as a racial project. Rather, 'blood' stands for a reservoir of cultural and historical memory, a tradition, a body of writing, and the potential incarnation that Deronda and other Jews may accept, or may decide to reject, depending on their relationship to it. Further, pointing to Judaism as an example, *Daniel Deronda* figures religion as ultimately a matter of election. Even as the Jewish people are figured as chosen, they are also understood as choosing as well. Eschewing racial determination, Eliot's portrayal of Judaic religiosity thus can serve as an example for Christianity to emulate.

Transcendence through separateness

As I have suggested, *Daniel Deronda* represents Judaism as a form of religious resistance to secularization. Although Eliot was a secularist, it is clear from the novel and 'The Modern Hep! Hep! Hep!' that she saw secularization as potentially destabilizing and corrupting of individual and national character under existing conditions. Thus, the accommodation and recommendation of religiosity in the novel may be understood in anthropological terms; religion provides a means for preserving the best cultural, moral, and social traits of a people and its members, and for passing them along to future generations. Religion serves in *Daniel Deronda* a function analogous to that of Matthew Arnold's 'culture' in *Culture and Anarchy* (1869). Arnold assigned to 'culture' the role of providing an ideal toward which the subject should strive, a means for getting the 'fresh and free play of the best thoughts upon his stock notions and habits'.[80] The novel rejects Arnold's notion of 'culture' for its failure to provide an agenda, for its lack of a mission. As is evident in the interior portrait of the early Deronda, culture provided knowledge, not of everything, but only of everything *about* everything:

> He was ceasing to care for knowledge – he had no ambition for practice – unless they could both be gathered up into one current with his emotions; and he dreaded, as if it were a dwelling-place of lost souls, that dead anatomy of culture which turns the universe into a mere ceaseless answer to queries, and knows, not everything, but everything else about everything – as if one should be ignorant of nothing concerning the scent of violets except the scent itself for which one had no nostril.[81]

In place of the amorphous and 'dead anatomy of culture' disparaged in the above passage, in *Daniel Deronda,* religion embodies the direction and purpose that 'culture' cannot provide. Religion is able to avail Deronda 'some external event, or inward light' to guide his actions.

Further, religion delivers transcendence, and on terms antithetical to those that Arnold had recommended in connection with culture.

As Bryan Cheyette points out, in *Culture and Anarchy*, Arnold outlined and endorsed a universalizing culture by which 'fixed racial differences between "Aryans" and "Semites"' would be transcended.[82] As opposed to Arnold, *Daniel Deronda* and the 'The Modern Hep! Hep! Hep!' represent religion as conserving precisely those differences that Arnold's culture would transcend, while at the same providing for a transcendence of its own. While in *Culture and Anarchy* Arnold may be seen as effectively recommending the overwriting of cultural and 'racial' differences in a higher-level assimilation under the rubric of 'culture', in *Daniel Deronda* religion preserved those very cultural traits that make a people distinct. In 'The Modern Hep! Hep! Hep!' Eliot's persona extols a 'religion founded on historic memories' and 'characteristic family affectionateness' and views them as the means by which the Jewish people – 'tortured, flogged upon, the corpus vile on which rage or wantonness vented themselves' – managed nevertheless to have 'escaped with less of abjectness, and less of hard hostility toward the nations whose hand has been against them'.[83] These cultural and 'historic memories', meanwhile, are not only necessary for Jewish survival but of immense importance to those beyond the cultural and religious group. As Mordecai says to his sister Mirah after receiving news that Deronda would soon return to London:

> 'Seest thou Mirah', he said once, after a long silence, 'the *Shema*, wherein we briefly confess the divine Unity, is the chief devotional exercise of the Hebrew; and this made our religion the fundamental religion for the whole world; for the divine Unity embraced as its consequence the ultimate unity of mankind. See, then – the nation which has been scoffed at for its separateness, has given a binding theory to the human race'.[84]

The separateness of the Jewish people has yielded the concept of a divine unity, which becomes the conceptual means by which the entirety of humanity may be unified. As such, religion provides the means for cultural transcendence, but such transcendence is possible only because of the preservation of cultural distinctiveness.

Furthermore, the unity of human kind through the differentiation of the parts and cultural transcendence allows for the conceptualization of the unity of the Supreme Being:

> Now, in complete unity a part possesses the whole as the whole possesses every part: and in this way human life is tending toward the

> image of the Supreme Unity: for as our life becomes more spiritual by capacity of thought, and joy therein, possession tends to become more universal, being independent of gross material contact; so that in a brief day the soul of a man may know in fuller volume the good which has been and is, nay, is to come, than all he could possess in a whole life where he had to follow the creeping paths of the senses.[85]

While Mordecai's logic is circular – the concept of 'the divine Unity' leads to the unification of the human race, which makes possible the conception of 'the Supreme Unity' – the point is that spirituality, which unites the human race, derives from the particularity of the group, upon which depends the imagination of its transcendence and the universality of Judaism. As Mordecai (and Eliot) see it, the separateness and integrity of the Jewish people are necessary preconditions for the nurturance of the idea of the 'divine Unity', which lays the foundation for cultural and religious transcendence and universal unity. The phrase 'separateness with communication', which Deronda adopts from his grandfather through his grandfather's friend Joseph Kalonymos, makes clear that in the novel the preservation of difference does not come at the cost of cultural isolation.[86]

Yet, the question remains: Why does Eliot insist upon the prior separateness and peculiarity of Jewry as a precondition of cultural transcendence and the unity of humanity (through the concept of the 'Supreme Unity')? The answer has to do with what Eliot saw as the genius of nationality and the singular importance of preserving such genius against the pressures of cosmopolitan dilution:

> The tendency of things is toward the quicker or slower fusion of races. It is impossible to arrest this tendency: all we can do is to moderate its course so as to hinder it from degrading the moral status of societies by a too rapid effacement of those national traditions and customs which are the language of the national genius – the deep suckers of healthy sentiment. Such moderating and guidance of inevitable movement is worthy of all effort. And it is in this sense that the modern insistance on the idea of Nationalities has value.[87]

And the particular genius of the Jewish people, as Eliot saw it, is 'the religion of a people whose ideas have determined the religion of half the world, and that the more cultivated half'.[88] This is why she chose to write so extensively about Anglo-Jewry and Judaism in *Daniel*

Deronda. And, as she wrote to Harriet Beecher Stowe on 29 October 1876, Eliot

> felt urged to treat Jews with such sympathy and understanding as my nature and knowledge could attain to ... But towards the Hebrews we western people who have been reared in Christianity, have a peculiar debt and, whether we acknowledge it or not, a peculiar thoroughness of fellowship in religious and moral sentiment.[89]

Thus, Eliot wished to augment the sympathy for and understanding of Jewish people among her Christian readers, to work to decrease anti-Semitism and to increase philo-Semitism. She also felt that Christians owed a debt of gratitude to their Jewish predecessors and contemporaries and shared a 'fellowship in religious and moral sentiment', a cultural transcendence, that Christians failed to acknowledge. More than this, however, in both *Daniel Deronda* and 'The Modern Hep! Hep! Hep!', Eliot suggested that Judaism could serve as an exemplar of the kind of religious and moral sentiment that was waning among other British subjects. Clearly, Eliot believed that the religion of Judaism had something particular and indispensable to offer the British nation and the world at large. As I have suggested above, this was a belief that Judaism was capable of a defense against the secularizing, cosmopolitanizing tendencies of the period. To what extent was such a belief warranted? In the following section, I briefly take up this question by examining two indices of the Jewish response to the secular forces in the late nineteenth century.

Judaism and 'secularization'

While scholars have examined the importance of 'the Jewish Question' to major social and political movements and issues, such as the Enlightenment, liberalism, socialism, queer theory, literature, and national identity,[90] scant attention has been paid to Judaism in the context of secularism, secularization, or secularity in Great Britain.[91] The mid-century 'crisis of faith' and the questions regarding secularism and secularization have generally been considered almost exclusively in connection with Christianity, without consideration of Judaism, or any other faith for that matter. In this section, I hope to begin redressing this remission by briefly discussing two potential flash points for inaugurating an extended discussion in this new direction: the Judaic response to the Higher Criticism, and the reaction of Anglo-Jewry to

Darwinism. While I will draw these two threads of discussion together, a much more extended study would include a treatment of any Jewish members of the Secularist movements, as well as a broader analysis of Jewish secular writing in the period, among other issues.

While the Anglo-Jewish response to Darwinism has received some recent attention (as discussed below), very little has been written about the nineteenth-century Anglo-Jewish reaction to the Higher Criticism. More, in fact, has been said about the uses made of Judaism by Christians to rebut the Higher Criticism. As Michael Scrivener has observed, in addition to the typical toleration of Jews coupled with hostility toward Judaism that characterized British 'semitism',[92] another discursive stream also subsisted beside it, the discourse that figured Judaism as providing protection against secularization:

> As the established religious certainties are under attack by science, rationalism, and Enlightenment, Christians strategically use Judaism – the Old Testament – to provide a secure foundation for the religion of the New Testament. Against the Higher Criticism of the Bible there are the ever refined, constantly revised prophetic readings of the older Testament to confirm the truths of the newer.[93]

Similarly, Cynthia Scheinberg has suggests that while the Anglo-Jews were a small minority in Britain, 'the figure of the Jew' was significant in the social and political imaginary. It was invoked by Christian interlocutors in response to Darwinism and the Higher Criticism, and when concepts of racial difference were codified within the discourse of social Darwinism, as well as in debates over the political emancipation of non-Anglicans, including Catholics and Dissenters (granted in 1828 with the Repeal of the Test and Corporation Acts) and Jews (granted with the passage of the Jews Relief Act in 1858).[94]

But what were the responses of actual Anglo-Jews to the Higher Criticism, and do they bear testimony to the belief held among some Christians about the efficacy of Judaism in resisting biblical interpretations believed corrosive of religiosity? No systematic study of the nineteenth-century Anglo-Jewish response to the Higher Criticism has been published; probably the best account is still David Englander's brief discussion within a longer article from 1988.[95] As Englander points out, the two major Anglo-Jewish nineteenth-century responses to the Higher Criticism were those by Dr. Abraham Benisch, the progressive editor of the *Jewish Chronicle*, and Solomon Schechter, who replied to Benisch. Benisch published a book entitled *Bishop Colenso's Objections to the*

Historical Character of the Pentateuch and the Book of Joshua in 1863, in response to Bishop Colenso's contribution to *Essays and Reviews* (1860). The lengthy essay, which originally appeared in the *Jewish Chronicle* in several weekly installments between 28 November 1862 and 27 February 1863, mounted a rigorous defense of the Hebrew Bible. While Benisch claimed that his *Chronicle* readers had received the series with overwhelming approbation after some of 'the author's co-religionists ... had been unsettled by the Bishop's arguments',[96] as Englander points out, Benisch's defenses were no real match for the many problems posed by the attacks of Bishop Colenzo and others, especially as they challenged 'the heterogeneous composition of the Pentateuch, the comparatively late date of the Levitical Legislation, and the post-exilic origin of certain Prophecies as well as of the Psalms'.[97] Schechter's reply, also originally appearing in the *Jewish Chronicle,* was included in his volume entitled *Studies in Judaism,* first published in 1896. Schechter argued that the best strategy for defending Judaism from the attacks of both 'simple meaning (Philology)' and 'Natural Science' was to 'shift the centre of gravity in Judaism and to place it in the secondary meaning, thus making religion independent of philology and all its dangerous consequences'.[98] Schechter's point was that language is ambiguous and double in meaning. It was only when Judaic religionists attempted to defend a narrow, simple meaning of the Pentateuch that the Higher Criticism or the natural sciences could do any damage to Judaism. As opposed to such a defense of Biblical Judaism, Schechter recommended a Talmudic recourse to 'Jewish Tradition, or, as it is commonly called, the Oral Law'. Tradition or Oral Law evolves and develops, and is embodied in the works of the Rabbis from the Middle Ages on, Rabbis whose interpretations represent the 'Secondary Meaning of the Scriptures'. Schechter identified this approach with the 'historical school', whose scholars study the 'post-biblical literature, not only elucidating its texts by means of new critical editions, dictionaries, and commentaries, but also trying to trace its origins and to pursue its history through its gradual development'.[99]

The historical school's relationship to Jewish Orthodoxy was analogous to that of the Tractarian movement – or Catholicism itself – to orthodox British Protestantism. In fact, Schechter paid homage to Catholicism and the Oxford Movement, and referred to the tradition that he represented as 'Catholic Israel'.[100] Like Catholicism, the historical school suggested that '[i]t is not the mere revealed Bible that is of first importance to the Jew, but the Bible as it repeats itself in history, in other words, as it is interpreted by Tradition'. Instead of resting on the authority of the Bible, 'the centre of authority is actually removed from the Bible and placed in some *living body*'. The living body was 'the collective conscience of Catholic Israel as embodied in the Universal Synagogue'.[101]

As Englander notes, Schechter's approach, although it apparently absolved Anglo-Jewish religionists from having to take seriously the work of contemporary Biblical criticism, was atypical in Britain. His influence was greater in the United States, where such an approach appealed to observant Jews, who deemed both Orthodox and Reformed Judaism unacceptable. Schechter's well-known epithet for the Higher Criticism – 'the higher anti-Semitism' – 'was symptomatic of the marginal position which the Higher Criticism occupied within the consciousness of Anglo-Jewry', according to Englander.[102] In short, Englander suggests that while such polemicists and scholars as Benisch and Schechter were occupied with the Higher Criticism, the 'mass of Jews' paid little or no attention to it. Thus, at least where the Higher Criticism is concerned, the 'crisis of faith', if it can be so-called, among the Jewish intelligentsia, resembled that of the Christian intellectuals and literary artists, only on a much smaller scale. That is, it affected a slight segment of a class, but apparently in the case of Anglo-Jews, this class segment fended off the Higher Criticism fairly well. As we have seen, Christian Britain also included a substantial group of artisanal and working-class intellectuals who had openly abandoned Christianity decades before the middle-class crisis of faith became notorious. Much more work should be done, then, to examine the effects of such intellectual movements as the Higher Criticism on Anglo-Jewish working-class subjects.

The response to Darwinism represents another dimension of 'secularization' in connection with Judaism and Anglo-Jewry. As Geoffrey Cantor and Marc Swetlitz point out in their path-breaking, singular study of Darwinism and the challenges it may have posed for the Jewish tradition, '[i]nnovative science, of which evolution is a paradigm example, usually forms a central element of the modernist worldview, and reactions to evolution are often symptomatic of a wider response to modernism, progress, and social change'.[103] While this seems to suggest that innovative science is necessarily secularizing, I take it to mean that paradigm-shifting science that already has been secularized, such as Darwinian evolutionary theory, has the potential to destabilize traditional worldviews, and thus has secularizing potential. In any case, the point here is that Darwinian science can serve as a test case for its potential to act on Anglo-Jewish Judaism in a secularizing direction.

Geoffrey Cantor has written what is perhaps the first treatment of the initial Anglo-Jewish response to Darwinism in the nineteenth century; I will summarize his findings and generalize from them in terms of 'secularization' and secularity. After laying the contextual background regarding the state of poor Jewish education and general lack of participation in science during the period ('Until the mid-1860s or possibly later, the community's contribution to science and literature was indeed minimal'.[104]), Cantor also notes that before mid-century, in Christian Britain, science had not only been, for the most part, deemed compatible with religion but also under the rubric of Natural Theology, it served to buttress religious conviction. (See Chapter 2.) From Paley's *Natural Theology* (1802) through the last of the Bridgewater Treatises in 1836, Natural Theology enjoyed what has widely been understood by historians of science as an 'Indian Summer' in the first half of the nineteenth century.[105] That is, until roughly the 1850s, when the scientific naturalists rose to prominence (see Chapter 4), science was believed to complement rather than contradict religious belief.

Perhaps because they never had relied on Natural Theology or arguments from design in the first place, or because Judaism had been figured as 'rational' and compatible with reason throughout its long history, the Anglo-Jewish respondents to Darwinian evolution generally reacted positively to it. Although exceptions certainly existed, for the most part, Anglo-Jewish commentators found little or no incompatibility between Darwinian evolution and Judaism. Instead, contemporary natural science was deemed compatible with Judaism and Judaism to be specially and favorably situated in relation to it. Again, many of these commentaries took place in the *Jewish Chronicle*, as its editors often engaged with science and evolutionary science in particular, as well as with scientific naturalism. One of the *Chronicle*'s longer term editors, Dr. Abraham Benisch, who was also 'one of the few accomplished Torah and Talmud scholars in Britain',[106] was one such commentator. Benisch found sanction for the centrality of reason in Biblical verses, and praised the Bishops of London for siding with science over contemporary evangelical Christians. Myers Davis, the editor of the *Jewish World*, a down-market penny and competitor of the *Chronicle*, adopted a more aggressive posture. Responding to John Tyndall's famous 'Belfast Address' in 1874, Myers apparently reveled in the difficulties that the scientific naturalists posed for Christianity, while suggesting that the attacks of scientific naturalism upon Christianity vindicated Judaism and Jews. Although he advised his readers to resist entering the fray between science and Christianity, as Cantor points out, 'Myers was clearly hoping that the tormentors of the Jewish community would be humiliated by losing their battle with Tyndall, Huxley, and the other scientific naturalists. Judaism, so often vilified by Christians, would emerge the more resilient religion'.[107]

Cantor cites several other examples of what he calls 'the Standard Anglo-Jewish Response to Science', which I have just characterized. Much like the treatment of the Higher Criticism, then, the standard Anglo-Jewish response to the challenges of Darwinism appears to have been generally accommodating and in some cases even triumphal, especially as compared with the problems encountered by Christian apologists. Cantor also notes exceptions to this standard response, which came from both conservative and progressive quarters. In particular, he notes one progressive dissenter, Alfred Gutteres Henriques, whom I have mentioned above. Henriques held that evolutionary science demonstrated the mutability of the natural world, while Judaism upheld its immutability. He held that scientific naturalism outmoded traditional Judaism and that the latter had to be radically recast so that it might correspond with the findings of science. As Cantor puts it, '[i]ndeed, although the central dogma of God's unity still remained intact, modern science undermined so many facets of traditional Judaism that Henriques doubted whether it could ever be revivified as a religion after Darwin's onslaught'.[108] However, mainstream commentators blithely ignored the challenges that progressive Anglo-Jews like Henriques found in Darwinian science and Cantor names the standard response as such for a reason.

Conclusion: *Daniel Deronda*, Judaism, secularity

In this chapter, I have examined 'the Jewish Question' in connection with 'secularization', a sorely neglected line of research in nineteenth-century British studies, and one which demands much more inquiry to do it justice. Judaism must be considered in any serious study of nineteenth-century British secularism. I have approached the subject through the fiction of George Eliot, and finally in the responses of some of Eliot's Anglo-Jewish contemporaries to factors generally understood as contributory to 'secularization'. Other viable and important lines of inquiry in this area lay untouched to date. But based on preliminary studies, we can tentatively conclude that, at least in the nineteenth century, as George Eliot suggested, mainstream Anglo-Judaism indeed was more resilient, not only to Darwinism and scientific naturalism, but also to the Higher Criticism, than its Christian counterparts – and thus to 'secularization' as such. Exceptions to this rule no doubt existed, and we should not treat Judaism as a monolithic block. But the greater exception seems to be Judaism itself – another exception to the standard secularization thesis.

I have maintained a critical stance toward 'secularization' purposefully, because my argument here and throughout this book has been that 'secularization' is not what it has been made out to be. *Daniel Deronda* and the Anglo-Jewish responses to both the Higher Criticism and evolutionary science demonstrate that Judaism complicates the picture of 'secularization' considerably, and further supports my argument regarding the character of secularity. Judaism adds another element to secularity's pluralism. It not only registers a resilience to factors deemed 'secularizing' but also, as the capacious novel *Daniel Deronda* attests, it adds a dimension to understanding to the religious-secular configuration of Britain. Judaism and Anglo-Jewry represented an Other, as Bryan Cheyette has argued, an Other that was at once marginal and central. As the uses made of Judaism by Christian religious interlocutors attest, Judaism certainly took up more space in the social imaginary than its Jewish numbers would suggest.

Epilogue: Secularism as Modern Secularity

In 1910, just four years after Holyoake's death, the Hastings *Encyclopaedia of Religion and Ethics* included an entry on Secularism, but one that fell under the heading of Atheism. Within the subheading of Secularism, the 1910 edition rather sloppily announced the equivalence of Holyoake's and Bradlaugh's Secularism on the grounds of atheism, and professed both to be mistaken and problematic because they had relied on negation rather than the positing of distinct values. While Holyoake, Bradlaugh, and company surely had reasons for their hostilities and vituperations, they were essentially locked in a position of denunciation from which nothing positive could emerge.[1] Thus the revision of Secularism was well underway and Holyoake's particular contribution, in fact his construction of Secularism itself, was effectively erased and overwritten, as the two currents of Secularism were conflated.

But by 1920, the same encyclopaedia offered a separate heading for Secularism, and a description quite at odds with the previous interpretation. While characterizing Secularism as 'negatively religious' – by which the author meant that Secularism undertook the functions of 'morality' and 'a theory of life ... without reference to a deity' – the entry aptly characterized Secularism (as founded by Holyoake) as agnostic with reference to metaphysical questions: 'Neither theism nor atheism enters into the secularist scheme, because neither is provable by experience'.[2] Declaring that Secularism had sprung from particular political, social and economic conditions, which no longer obtained, the author pronounced organized Secularism defunct and unlikely to be resuscitated. Yet, he continued, Secularism should be evaluated in terms of its philosophical value, rather than strictly in terms of its organizational viability: 'The question', he argued, 'is rather whether its spirit and principles are destined to continue in being'.[3]

Calling attention to the two streams of Secularism, the 1920 entry, which did a much better job than the earlier entry to correctly position Secularism as a movement and creed based on positive principles and the eschewal of the metaphysical, continued by challenging the theoretical coherence of Holyoake's version and extolling the greater consistency and worldly impact of Bradlaugh's atheistic and anti-religious variety. The problem with Holyoake's Secularism, it argued, was that it claimed to ignore what cannot be ignored by any self-consistent and independent system:

> The attempt to ignore rather than deny religion is impractical, because religion embraces both secular and spiritual concerns. Religion denies the secular *conception* of life, and that conception cannot establish itself without defeating the claim of religion to control life. It is an impossible proposition to maintain that there may be a God, but that He does not concern material existence.[4]

While Holyoake never suggested that God 'does not concern material existence', rather only that the Secularist need not (although she may) concern herself with God, the author of the entry made some salient points. He rightly pointed out that religion not only lays claim to religious life and concerns, to the otherworldly, but it also stakes its claims on the secular as well, to 'this world'. Since religion denies the secular conception of life, or life construed as strictly a secular matter, the secular conception can only be positively asserted by negating the religious conception. For this reason, Bradlaugh's Secularism was both the more coherent and the more successful type. Bradlaugh was correct, theoretically and practically, the entry's author continues, to attack that which barred Secularism's claims over secular life. For his part, Holyoake had inadvertently yielded the ground of the secular to theology. This mistake explained the relative weakness of his position, and the greater the success of the negative strain. Further, it explained why Secularism as a whole only found firm footing during periods of religious repression and persecution, and during the heightened opposition between science and religion.[5]

While it is certainly debatable whether or not a philosophical system can subsist without addressing the question of deity (either positively or negatively), the more important point here for my purposes has to do with the tensions that the author of this encyclopedia entry registers regarding Holyoake's Secularism. With Secularism, Holyoake envisioned a broad tent movement based on an agnostic ecumenism within which both secular and religious elements and persons might subsist and cooperate. Secularism, as I have illustrated, was not meant to represent the antithesis of religion. It was supposed to represent the cooperation of religious and non-religious members for secular ends.

Yet another strain runs through Holyoake's writing on Secularism. Holyoake also proffered Secularism as a replacement for the theological within religion, as a moral system based on secular, materialist premises, especially intended for those who rejected theology and theism outright. As late as 1871, in *The Principles of Secularism Illustrated*, Holyoake declared unequivocally:

> Secularism is a series of principles intended for the guidance of those who find theology indefinite, or inadequate, or deem it unreliable. It *replaces* theology, which mainly regards life as a sinful necessity, as a scene of tribulation through which we pass to a better world.[6]

Likewise, under the overarching umbrella of Secularism, where together theists and atheists would supposedly cooperate for secular ends, Holyoake apparently attempted to smuggle in the *desideratum* for Secularist hegemony, a mandate to overcome theology and theism altogether – if not by destroying them in a head-on confrontation, then by a process of attrition by virtue of Secularism's greater credibility as a system. Nevertheless, Holyoake continued in an apparent attempt to placate those (theists) whom he may have just alienated: 'Secularism rejoices in this life, and regards it as the sphere of those duties which educate men to fitness for any future and better life, should such transpire'. Thus, in the span of a few short sentences, Holyoake straddled the fence between the denial of the otherworldly and the accession to its possibility, and likewise the supersession and inclusion of theology.

Treating Holyoake charitably and acknowledging the practical concerns and difficulties that he felt compelled to manage – such as the involvement of both theists and the holdovers from the earlier, more trenchant freethought movement – we can read him as negotiating the difficulties and differences of a varied constituency, caught in the middle and attempting to construe a pragmatic, adaptable Secularism amidst antagonistic forces. Or, we may view Holyoake's program and policies as mistaken – as simply self-contradictory.

I want to suggest that the Secularism that Holyoake struggled to elaborate and build, while significant as a comprehensive philosophy and social and political movement, is less important as such than as the expression of a *condition* – at the moment of its inception – an epitome that embraced such antinomies as philosophers and activists might find utterly incompatible. Thus, we need not work out the tensions that subsisted within Holyoake's Secularism, or dismiss it on the basis of its supposed philosophical incoherence or practical failures. Holyoake's Secularism best represents a theoretical and organizational expression of the condition of the secular-religious moment at mid-century. Secularism's greatest significance lies in its historicity, its inaugural expression, at a crucial juncture, of modern secularity or the post-secular condition.

This understanding allows Secularism to serve as a hermeneutic key for making sense of developments in the period – for understanding that the emergence of the secular often has been coupled with re-enchantment, that the 'crisis of faith' has been accompanied by a 'crisis of doubt', that de-conversions sometimes have led to reconversions and conversions to de-conversions, that scientific naturalism emerged successfully as a form of agnosticism rather than as atheism or a hard naturalism, that the parrying between belief and unbelief and science and religion continued after the emergence of the secular in science, and so on. This condition of secularity is the result of 'secularization', but secularization understood as leading to this very conditionality, rather than understood as the ultimate elimination of religion or religiosity.

Understanding Secularism as such, as the emergence of modern secularity, allows us to see that secularity embraces both the persistence and the conditioning of religion as well as such elements as 'hard secularism' – as seen in the positions of Richard Carlile and Charles Bradlaugh and its persistence within the Secularist movement. Secularism or secularity should be understood as an *optative* condition that expresses a secular-religious pluralism and the effective indeterminacy with reference to its options. As I have suggested in Chapter 3, mid-century Secularism should be understood not as a failed mid-century movement, nor, as having nothing to do with 'secularization' broadly construed. Secularism is a monumental expression of secularization as it arrives at modern secularity or the post-secular condition.

Whether secularity as it emerges in the mid-nineteenth century endures beyond that historical threshold moment has been beyond the scope of this study to explore. However, I submit that this understanding of secularity as such depends on its experiential possibility in our own time. That is, secularity represents our own naïve 'background condition', and without our experience of it we could neither trace its emergence nor contrast it with its historical precedent in the holistic Christian cosmology that it replaced. Again, this apparent persistence only makes the arrival of secularity with Secularism that much more important.

Finally, this study, I hope, has suggested further lines of inquiry for exploring secularity in the nineteenth century and beyond. I have hinted at some of these threads in Chapter 6 in connection with 'secularization' and Judaism. Other chapters may also serve as examples of the kind of episodes that secularity has involved. I look forward to pursuing further contributions along these lines, as well as seeing them pursued by others.

Notes

Introduction: Secularity or the Post-Secular Condition

1. For a discussion of the 'death of God' discourse, see, for example, J. W. Robbins 'Introduction: After the Death of God', in J. D. Caputo, G. Vattimo, and J. W. Robbins (2009) *After the Death of God* (New York, NY: Columbia University Press), esp. at pp. 1–10.
2. Anon. (25 February 1968) 'A Bleak Outlook Is Seen for Religion', *New York Times*, 3.
3. Once an important secularization theorist, Berger reversed his long-standing position on secularization in P. L. Berger (1999) *The Desecularization of the World: Resurgent Religion and World Politics* (Washington, DC: Ethics and Public Policy Center).
4. For a summary of the secularization debates, see R. Warner (2010) *Secularization and Its Discontents* (London; New York: Continuum); D. V. A. Olson and W. H. Swatos (2000) *The Secularization Debate* (Lanham, MD: Rowman & Littlefield Publishers); S. Bruce (ed.) (1992) *Religion and Modernization: Sociologists and Historians Debate the Secularization Thesis* (Oxford: Oxford University Press); I. Katznelson and G. Stedman Jones (eds) (2010) *Religion and the Political Imagination* (Cambridge: Cambridge University Press). For a call to abandon the notion of secularization, at least where historians are concerned, see D. Nash (2004) 'Reconnecting Religion with Social and Cultural History: Secularization's Failure as a Master Narrative', *Cultural and Social History* 1, 302–25.
5. T. Asad (2003) *Formations of the Secular: Christianity, Islam, Modernity* (Stanford, CA: Stanford University Press), p. 46.
6. J. Habermas (2008) 'Notes on Post-Secular Society', *NPQ: New Perspectives Quarterly* 25.4, 17–29. In March of 2007, Habermas delivered his now famous lecture on 'post-secularism' at the Nexus Institute of the University of Tilberg, Netherlands. Habermas pointed to three factors that characterize modern social orders as post-secular: 1) the broad perception that many global conflicts hinge on religious strife and the changes in public consciousness and weakening of confidence in the dominance of a secular outlook that such acknowledgement accedes; 2) the increased importance of religion in various public spheres; and 3) the growing presence in Europe and elsewhere of immigrant or 'guest workers' and refugees with traditional cultural backgrounds.
7. For the persistence of the secularization thesis, see S. Bruce (2002) *God Is Dead: Secularization in the West* (Malden, MA: Blackwell Pub). For a significant revision, see P. Norris and R. Inglehart (2011) *Sacred and Secular: Religion and Politics Worldwide* (Cambridge: Cambridge University Press). Norris and Inglehart advance the 'existential security hypothesis' as the explanation for secularization or the lack thereof. According to this thesis, as populations

become relatively secure economically and otherwise, religiosity tends to decline. The denizens of post-industrial societies are demonstrably less religious than those living in agricultural or industrial economies. Meanwhile, although secularization is increasing as regions become post-industrial, religious populations are growing relative to secular ones, owing to the fact that in traditional societies the birthrate is much higher than in secular societies.

8. Asad questions the very existence of 'modernity' as such.
9. C. LaPorte (2013) 'Victorian Literature, Religion, and Secularization', *Literature Compass* 10.3, 277–87, at 277.
10. See, for example, O. Chadwick (1975) *The Secularization of the European Mind in the Nineteenth Century* (Cambridge: Cambridge University Press); H. Cox (1965) *The Secular City: Secularization and Urbanization in Theological Perspective* (London: Macmillan); A. D. Gilbert (1980) *The Making of Post-Christian Britain: A History of the Secularization of Modern Society* (London: Longman); D. Martin (1978) *A General Theory of Secularization* (Oxford: Blackwell); M. H. Abrams (1971) *Natural Supernaturalism: Tradition and Revolution in Romantic Literature* (New York, NY and London: WW Norton); and B. Willey (1956) *More Nineteenth-Century Studies: A Group of Honest Doubters* (New York: Harper & Row).
11. For criticism and revision of Romanticism and/as secularization, see C. Jager (2014) *Unquiet Things: Secularism in the Romantic Age* (Philadelphia: University of Pennsylvania Press); C. Jager (2008) 'Romanticism/ Secularization/ Secularism', *Literature Compass* 5.4, 791–806; and C. Jager (2007) *The Book of God: Secularization and Design in the Romantic Era* (Philadelphia: University of Pennsylvania Press).
12. B. Holsinger (2006) 'Literary History and the Religious Turn', *ELN* 44.1, 1.
13. T. Larsen (2006) *Crisis of Doubt: Honest Faith in Nineteenth-Century England* (Oxford: Oxford University Press). See also D. Nash (2011) 'Reassessing the "Crisis of Faith" in the Victorian Age: Eclecticism and the Spirit of Moral Inquiry', *Journal of Victorian Culture* 16.1, 65–82; and F. M. Turner (2011) 'Christian Sources of the Secular', *Britain and the World* 4.1, 5–39.
14. See C. Brown (2009) *The Death of Christian Britain: Understanding Secularisation, 1800–2000*, 2nd ed. (London: Routledge), p. 10. Note that *Nineteenth-Century British Secularism* does not suggest that the timing or gradient of secularization should be revised, but rather that the end of secularization should be reconsidered and reconstructed through a new notion of 'secularity'.
15. C. Brown, *The Death of Christian Britain*, p. 11.
16. For the argument that secularization never happened, see A. Morozov (2008) 'Has the Postsecular Age Begun?', *Religion, State & Society* 36.1, 39–44.
17. D. Nash, 'Reassessing the "Crisis of Faith"', p. 70.
18. C. Taylor (2007) *A Secular Age* (Cambridge, MA: Belknap Press of Harvard University Press).
19. D. Nash (2013) *Christian Ideals in British Culture: Stories of Belief in the Twentieth Century* (Houndmills, Basingstoke, Hampshire; New York: Palgrave Macmillan), p. 15.
20. The major social historian of Secularism is E. Royle (1974) *Victorian Infidels: The Origins of the British Secularist Movement, 1791–1866* (Manchester: University of Manchester Press; Totowa, NJ: Rowman & Littlefield); and E. Royle (1980) *Radicals, Secularists, and Republicans: Popular Freethought in Britain, 1866–1915*

(Manchester: Manchester University Press; Totowa, NJ: Rowman & Littlefield). See also D. Nash (1992) *Secularism, Art, and Freedom* (Leicester: Leicester University Press); L. Schwartz (2013) *Infidel Feminism: Secularism, Religion and Women's Emancipation, England 1830–1914* (Manchester, UK: Manchester University Press). For general studies of unbelief, see S. Budd (1977) *Varieties of Unbelief: Atheists and Agnostics in English Society, 1850–1960* (London: Heinemann Educational Books); D. Berman (1988) *A History of Atheism in Britain: From Hobbes to Russell* (London and New York: Croom Helm); and S. A. Mullen (1987) *Organized Freethought: The Religion of Unbelief in Victorian England* (New York: Garland).

21. J. Marsh (1998) *Word Crimes: Blasphemy, Culture, and Literature in Nineteenth-Century England* (Chicago: University of Chicago Press); and M. Rectenwald (2013) 'Secularism', in M. Harris (ed.) *George Eliot in Context* (Cambridge: Cambridge University Press), pp. 271–78.
22. L. Schwartz (2013) *Infidel Feminism: Secularism, Religion and Women's Emancipation, England 1830–1914* (Manchester, UK: Manchester University Press).
23. M. Rectenwald (2013) 'Secularism and the Cultures of Nineteenth-Century Scientific Naturalism', *British Journal for the History of Science* 46.2, 231–54.
24. Anon. (17 March 1800) 'The Life of John Bunyan', *The Missionary Magazine*, No. 46, 97.
25. T. Asad, *Formations of the Secular*.
26. T. Asad, *Formations of the Secular*, p. 25.
27. B. Nongbri (2013) *Before Religion: A History of a Modern Concept* (New Haven: Yale University Press). This assertion is analogous to argument made by Butler regarding belief and unbelief. 'Neither belief nor unbelief is an origin', Butler states, suggesting that belief and unbelief constitute a discursive pair, and that each element in the pair becomes the source of its opposite. Unbelief or 'loss of faith' in the mid-nineteenth century is generative of a subsequent belief or faith, and so on. See L. S. Butler (1990) *Victorian Doubt: Literary and Cultural Discourses* (New York: Harvester Wheatsheaf), pp. 1–8, at 3.
28. D. Nash, *Christian Ideals in British Culture*, p. 17.
29. [W. H. Ashurst] (25 June 1851) 'On the Word "Atheist"', *Reasoner* 11.6, 88.
30. When referring to Secularism proper as originated and developed by Holyoake and company, I will use an uppercase 'S'. For the more general sense of the term, secularism in the lower case will be used.
31. N. Vance (2013) *Bible and Novel: Narrative Authority and the Death of God* (Oxford: Oxford University Press), p. 7.
32. V. Geoghegan (2000) 'Religious Narrative, Post-secularism and Utopia', *Critical Review of International Social and Political Philosophy* 3.2–3, 205–24, at 206.
33. A. Morozov, 'Has the Postsecular Age Begun?', p. 41.
34. C. Taylor, *A Secular Age*, p. 3.
35. M. Warner, J. VanAntwerpen, and C. J. Calhoun (eds) (2010) *Varieties of Secularism in a Secular Age* (Cambridge, MA: Harvard University Press), p. 22: '[B]ecause [Taylor's] third sense of the secular comprehends precisely those forms of religiosity that are now most widely mobilized, resurgence of religion is not evidence of a new post-secular dispensation'.
36. N. Vance, *Bible and Novel: Narrative Authority and the Death of God*, p. 17.

37. R. Bhargava (2015) 'We (In India) Have Always Been Post-Secular', in M. Rectenwald, R. Almeida, and G. Levine (eds) *Global Secularisms in a Post-Secular Age* (Boston and Berlin: De Gruyter), pp. 109–35.
38. D. Nash, *Christian Ideals in British Culture*, p. 6.
39. R. M. Young (1985) *Darwin's Metaphor: Nature's Place in Victorian Culture* (Cambridge: Cambridge University Press). Young sees the Darwinian 'revolution' as a demarcation debate within Natural Theology.
40. C. LaPorte (2013) 'Victorian Literature, Religion, and Secularization', p. 283.

1 Carlyle and Carlile: Late Romantic Skepticism and Early Radical Freethought

1. C. Taylor (2007) *A Secular Age* (Cambridge, MA: Belknap Press of Harvard University Press).
2. C. Taylor, *A Secular Age*, p. 351.
3. J. Habermas (2008) 'Notes on Post-Secular Society', *NPQ: New Perspectives Quarterly* 25.4, 17–29, at 27. Habermas sees 'hard naturalism' as a scientistic secularism, a secularism that attempts to justify itself and its polemical stance with regards to religion on scientific terms, despite the fact that the claims of the hard naturalist 'cannot be scientifically justified'.

   Ironically, and perhaps significantly for our narrative, Carlile would eventually espouse a highly unconventional and peculiar 'allegorical' or 'rational' Christianity, which figured Christianity as an allegorical system representing moral and physical entities that spanned numerous religions, as well as forms of paganism.
4. T. Larsen (2006) *Crisis of Doubt: Honest Faith in Nineteenth-Century England* (Oxford: Oxford University Press).
5. D. Nash (2011) 'Reassessing the "Crisis of Faith" in the Victorian Age: Eclecticism and the Spirit of Moral Inquiry', *Journal of Victorian Culture* 16.1, 65–82, at 70.
6. M. H. Abrams (1971) *Natural Supernaturalism: Tradition and Revolution in Romantic Literature* (New York, NY and London: WW Norton), pp. 12 and 68.
7. C. Jager (2008) 'Romanticism/Secularization/Secularism', *Literature Compass* 5.4, 791–806. For examples of such work see R. M. Ryan (1997) *The Romantic Reformation: Religious Politics in English Literature, 1789–1824* (Cambridge: Cambridge University Press); and W. A. Ulmer (2001) *The Christian Wordsworth, 1798–1805* (Albany: State University of New York Press). While aiming to correct Abrams's thesis, these studies, Jager suggests, have merely confirmed it. By focusing on the religious, they support Abrams's point, which was that the secular transforms and sometimes preserves religion. See C. Jager, 'Romanticism/Secularization/Secularism', pp. 794–96.
8. C. Jager, 'Romanticism/Secularization/Secularism', pp. 795 and 800. The studies Jager has in mind include M. Priestman (1999) *Romantic Atheism: Poetry and Freethought, 1780–1830* (Cambridge: Cambridge University Press); N. Roe (1997) *John Keats and the Culture of Dissent* (Oxford: Clarendon Press); D. E. White (2006) *Early Romanticism and Religious Dissent* (Cambridge, UK: Cambridge University Press); and M. Canuel (2002) *Religion, Toleration, and British Writing, 1790–1830* (Cambridge, UK: Cambridge University Press).

Only Canuel's study recognizes the institutional dimensions of secularism, according to Jager.

9. F. M. Turner (2011) 'Christian Sources of the Secular', *Britain and the World* 4.1, 5–39.
10. P. L. Berger (1967) *The Sacred Canopy: Elements of a Sociological Theory of Religion* (Garden City, NY: Doubleday).
11. F. M. Turner, 'Christian Sources of the Secular', p. 7.
12. D. B. Hindmarsh (2005) *The Evangelical Conversion Narrative: Spiritual Autobiography in Early Modern England* (Oxford: Oxford University Press). Meanwhile, the evangelical conversion reflects the narrative of pilgrimage and faith discovery found in earlier texts, especially *Pilgrim's Progress*. Thus, writing should be considered as both productive of such conversions, and reflective of them.
13. B. V. Qualls (1982) *The Secular Pilgrims of Victorian Fiction: The Novel as Book of Life* (Cambridge, UK: Cambridge University Press), esp. pp. 1–29.
14. According to Taylor, disenchantment results from religious reformism and precedes (and conditions) Enlightenment rationalism.
15. J. Landy and M. Saler (2009) 'Introduction: The Varieties of Modern Enchantment', in J. Landy and M. T. Saler (eds) *The Re-Enchantment of the World: Secular Magic in a Rational Age* (Stanford, CA: Stanford University Press), pp. 2–3.
16. Here I am borrowing from S. During (2013) 'George Eliot and Secularism', in A. Anderson and H. E. Shaw (eds) *A Companion to George Eliot* (Chichester; Malden, MA: Wiley-Blackwell), pp. 428–41, at 430, emphasis added. S. During refers to George Eliot's fiction in terms of a 'secular de-secularization of the secular'. Since the second 'secular' maintains the secular as a category, the doubling of 'secular' seems to me to be unnecessary, especially in the case of Carlyle, where the emphasis should be on the *desecularization* of the secular.
17. As many scholars have shown, *Sartor* involves the reader in editorial problems of her own. First is the problem of translation. This problem has been specifically treated by T. Toremans (2011) '*Sartor Resartus* and the Rhetoric of Translation', *Translation and Literature* 20, 61–78. Second is the question of the Editor's willful reading of the notes and letters of Teufelsdröckh, in particular his attempts to 'convert' the Clothes Philosophy and the philosopher's biography into a standard Christian conversion narrative. Given, as B. V. Qualls has shown, that the Editor eventually experiences a 'Conversion' of his own – to natural supernaturalism – I do not treat the problem of the difference between Editor's narrative and the posited underlying Clothes Philosophy and biographical notes provided by Teufelsdröckh. See B. V. Qualls, *Secular Pilgrims*, p. 28.
18. T. Carlyle and A. MacMechan (1896) *Sartor Resartus* (Boston and London: Ginn & Co.), p. 3.
19. T. Carlyle, *Sartor*, p. 29.
20. The dilemma faced by the Editor is similar to be that discussed by M. Foucault in 'What is an Author?' How do we know what constitutes the 'work' of an author? Do we include a laundry list found among her papers, for example? Why or why not?
21. T. Carlyle, *Sartor*, p. 45.
22. T. Carlyle, *Sartor*, p. 70.

23. Notably, as we shall see later in this chapter, R. Carlile celebrates some of these very men of science and for the very reason that *Sartor* criticizes them.
24. T. Carlyle, *Sartor*, p. 65, emphasis in original.
25. T. Carlyle, *Sartor*, p. 103.
26. T. Carlyle, *Sartor*, p. 60.
27. As F. M. Turner sees it in 'Christian Sources of the Secular', p. 10, 'clothes for Carlyle represented Kant's phenomenal realm while the inner spiritual reality of existence hidden beneath the clothing was Kant's noumenal realm'. This is certainly defensible, for as Carlyle figures it, the inner spiritual reality is finally inaccessible in the text. The only qualification I would make is that for Carlyle, the *noumena* finally becomes accessible in 'Natural Supernaturalism'.
28. T. Carlyle, *Sartor*, p. 194, emphasis in original.
29. 'Although today we think of *secular* as something that is contrasted with *religion*, the root notion of the term is something juxtaposed not to religion but to eternity. It derives from the same Latin etymological root – *saeculum* – as the French word *siècle*, meaning "century" or "age"'. C. Calhoun, M. Juergensmeyer, and J. Vanantwerpen (2011) 'Introduction', in C. Calhoun, M. Juergensmeyer, and J. Vanantwerpen (eds) *Rethinking Secularism* (New York: Oxford University Press), p. 8.
30. T. Carlyle, *Sartor*, p. 203.
31. T. Carlyle, *Sartor*, p. 90.
32. B. V. Qualls, *Secular Pilgrims*, p. 7.
33. T. Carlyle, *Sartor*, p. 109.
34. T. Carlyle, *Sartor*, p. 156.
35. T. Carlyle, *Sartor*, p. 201.
36. J. H. Miller (1989) '"Hieroglyphical Truth", *Sartor Resartus*: Carlyle and the Language of Parable', in J. Clubbe and J. Meckier (eds) *Victorian Perspectives: Six Essays* (Newark: University of Delaware Press), pp. 1–20, at 7, emphasis in original.
37. J. H. Miller, '"Hieroglyphical Truth"', p. 7, emphasis in original.
38. T. Toremans (2012) '"One Step from Politics": *Sartor Resartus* and Aesthetic Ideology', *Studies in the Literary Imagination* 45.1, 23–41, at 39. Incidentally, Toremans argues that in later works, Carlyle engages in no such referential circularity and does succumb to the totalitarian temptation.
39. T. Carlyle, *Sartor*, pp. 31 and 183.
40. T. Carlyle, *Sartor*, p. 167.
41. T. Carlyle, *Sartor*, p. 200.
42. T. Carlyle, *Sartor*, p. 231.
43. N. Paige writes in the context of late seventeenth- and eighteenth-century French debunking literature that '[d]isenchantment can be described as the process by which discourse that purports to describe the real world is stripped of its referential credentials ...' In *Sartor*, on the other hand, not only with disenchantment but also in the original enchantment the referential character of discourse is itself already occluded. In both cases, however, it is disenchantment that allows for the ultimate reestablishment of referentiality. In *Sartor*, this comes with natural supernaturalism. See N. Paige (2009) 'Permanent Re-Enchantments: On Some Literary Uses of the Supernatural from Early Empiricism to Modern Aesthetics', in J. Landy and M. Saler (eds)

The Re-Enchantment of the World: Secular Magic in a Rational Age (Washington, DC: The George Washington University Institute for Ethnographic Research), pp. 160–81, at p. 167.
44. T. Carlyle, *Sartor*, p. 169.
45. T. Carlyle, *Sartor*, p. 165.
46. J. Landy and M. Saler, 'Introduction: The Varieties of Modern Enchantment', pp. 1–2.
47. T. Carlyle, *Sartor*, p. 23.
48. T. Carlyle, *Sartor*, p. 231.
49. T. Carlyle, *Sartor*, p. 233.
50. T. Carlyle, *Sartor*, p. 241.
51. T. Carlyle, *Sartor*, p. 234.
52. T. Carlyle, *Sartor*, p. 235.
53. T. Carlyle, *Sartor*, p. 245.
54. H. B. Bonner and J. M. Robertson (1895) *Charles Bradlaugh: A Record of His Life and Work* (London: T.F. Unwin), pp. 19–20.
55. D. Nash suggests that some contemporary sociologists and others who maintain the standard secularization thesis, regardless of empirical evidence and theoretical disputation to the contrary, in fact hold such a belief. See D. Nash (2004) 'Reconnecting Religion with Social and Cultural History: Secularization's Failure as a Master Narrative', *Cultural and Social History* 1, 302–25.
56. R. Carlile (1821) *An Address to Men of Science: Calling Upon Them to Stand Forward and Vindicate the Truth from the Foul Grasp and Persecution of Superstition* (London: Printed and published by R. Carlile), p. 18.
57. For a biographical treatment of R. Carlile, see J. Wiener (1983) *Radicalism and Freethought in Nineteenth-Century Britain: The Life of Richard Carlile* (Westport, CT: Greenwood Press). For an extensive treatment of Carlile in connection with popular radicalism of the period, see J. Epstein (1994) *Radical Expression: Political Language, Ritual, and Symbol in England, 1790–1850* (New York: Oxford University Press). Epstein is particularly good at considering the rhetoric of Carlile's rationalist republicanism. For an intensive study of the rhetoric of radicalism, including that of Carlile, see K. Gilmartin (1996) *Print Politics: The Press and Radical Opposition in Early Nineteenth-Century England*, Cambridge Studies in Romanticism 21 (Cambridge and New York: Cambridge University Press).
58. J. Epstein, *Radical Expression*, p. 106.
59. M. L. Bush and R. Carlile (1998) *What Is Love?: Richard Carlile's Philosophy of Sex* (London: Verso), p. vii.
60. *Republican* (1819) 1, 330.
61. J. H. Wiener, *Radicalism and Freethought*, pp. 110–11. Carlile's sentence ended in 1823 but he was immediately arrested and returned to prison for not paying the fine of £1,500, so the process continued until he was eventually released on 25 November 1825.
62. J. Epstein, *Radical Expression*, p. 127.
63. J. Epstein, *Radical Expression*, p. 127; M. Butler (1996) 'Frankenstein and Radical Science', *Frankenstein: The 1818 Text, Contexts, Criticism* (New York and London: Norton), pp. 302–13; J. H. Wiener, *Radicalism and Freethought*, pp. 110–11.
64. The public debate in the late teens involved the two Professors at the Royal Society of Surgeons, J. Abernethy (1817) *Physiological Lectures, Exhibiting*

a General View of Mr. Hunter's Physiology (London: Longman, Hurst, Rees, Orme and Brown); and Sir W. Lawrence (1816) *An Introduction to Comparative Anatomy and Physiology; Being the Two Introductory Lectures Delivered at the Royal College of Surgeons, on the 21st and 25th of March 1816* (London: Callow) The latter was published in 1828 but written twelve years earlier. It was joined by T. Rennell, 'Christian Advocate' in the University of Cambridge, and G. D'Oyley, reviewing the debate in 1819 for the *Quarterly Review* 22, 1–35. S. T. Coleridge was a late entrant with the posthumously published S. T. Coleridge and S. B. Watson, M. D. (ed.) (1848) *Hints Towards the Formation of a More Comprehensive Theory of Life* (London: John Churchill), which had been written during the time of the debate, in 1816. Coleridge was a sophisticated vitalist.

65. J. H. Wiener, *Radicalism and Freethought*, pp. 110–12; M. Butler, 'Frankenstein and Radical Science', p. 312, note 6.
66. S. Shapin and A. Thackray (1974) 'Natural Knowledge in Cultural Context: The Manchester Model', *American Historical Review* 79, 672–709, at 678.
67. J. Epstein, *Radical Expression*, p. 124.
68. P. Trollander (1997) 'Politics of the Episteme: The Collapse of the Discourse of General Nature and the Reaction to the French Revolution', in L. Plummer (ed.) *The French Revolution Debate in English Literature and Culture* (Westport, CT: Greenwood Press), pp. 115–17; G. Kelly (1978) 'Intellectual Physicks: Necessity and the English Jacobin Novel', *Etudes Anglaises* 1, 161–75; D. M. Knight (1992) *Humphry Davy: Science & Power*, Blackwell science biographies (Oxford, UK: Blackwell), p. 35.
69. J. P. Klancher (2013) *Transfiguring the Arts and Sciences: Knowledge and Cultural Institutions in the Romantic Age* (Cambridge, UK: Cambridge University Press), p. 129.
70. J. P. Klancher, *Transfiguring the Arts and Sciences*, p. 128. For an account of the Science Wars, see A. Ross (1996) *Science Wars* (Durham, NC: Duke University Press), esp. at pp. 1–15.
71. For a discussion of the 'theory-ladenness of observation' see K. Knorr-Cetina and M. Mulkay (1983) *Science Observed: Perspectives on the Social Study of Science* (London: Sage Publications), pp. 1–17, esp. at 4–6.
72. See D. L. Robbins (1974) 'A Radical Alternative to Paternalism: Voluntary Association and the Popular Enlightenment in England and France, 1800–1840', unpublished doctoral dissertation, pp. 23–64.
73. As M. Priestman observes in *Romantic Atheism*, p. 10, the term 'Priestcraft' was suitably nebulous. Along with 'superstition', 'Priestcraft' always indicated Catholicism, but could also refer to the Church of England, or to all of Christianity.
74. T. Carlile, *An Address*, p. 26.
75. T. Carlile, *An Address*, p. 22.
76. T. Carlile, *An Address*, p. 47.
77. T. Carlile, *An Address*, pp. 30, 26.
78. J. P. Klancher, *Transfiguring the Arts and Sciences*, p. 126.
79. T. Carlile, *An Address*, p. 19.
80. T. Carlile, *An Address*, p. 5.
81. D. M. Knight, *Humphry Davy*, p. 5.
82. T. Carlile, *An Address*, p. 25.

83. T. Carlile, *An Address*, p. 33.
84. T. Carlile, *An Address*, p. 37.
85. T. Carlile, *An Address*, p. 4.
86. 'A Country Gentleman' (1826) *The Consequences of a Scientific Education to the Working Classes of this Country Pointed out; and the Theories of Mr. Brougham on that Subject Confuted; in a Letter to the Marquess of Lansdown. By a Country Gentleman* (London: Printed for T. Cadell), at pp. 54 and 71. Tracts by conservative opponents also included E. W. Grinfield (1825) *A Reply to Mr. Brougham's 'Practical Observations Upon the Education of the People; Addressed to the Working Classes and their Employers'* (London: Printed for C. and J. Rivington); and Rev. G. Wright (1826) *Mischiefs Exposed, a Letter Addressed to Henry Brougham, Esp., M.P., Shewing the Inutility, Absurdity, and Impolicy of the Scheme Developed in His 'Practical Observations' for Teaching Mechanics and Labourers the Knowledge of Chemistry, Mathematics, Party and General Politics, &c. &c.* Conservative tracts usually fell into what D. L. Robbins, in 'A Radical Alternative to Paternalism', p. 36, has termed the 'paternalistic' system of popular education: 'The paternalistic system belonged to an essentially unitary theocratic tradition, characterized by (1) a dominant faith in the power for good of moral and religious orthodoxy; (2) supervision of this orthodoxy by a monopolistic closed corporation; and (3) a stress on the virtues of reconciliation to social immobility'. Robbins sees Coleridge's 'Clerisy' as mapping closely onto prong 2.
87. For an excellent discussion of the 'the atheism debates', see M. Priestman, *Romantic Atheism*, pp. 12–43. Priestman's discussion is largely derivative of D. Berman (1988) *A History of Atheism in Britain: From Hobbes to Russell* (London: Croom Helm), pp. 110–33.
88. C. Calhoun (2010) 'The Public Sphere in the Field of Power', *Social Science History* 34.3, 301–35, at 316–28.
89. The best treatment of the conflict thesis is by J. H. Brooke (1991) *Science and Religion: Some Historical Perspectives* (Cambridge: Cambridge University Press), esp. at pp. 33–42. See also the recent elaborations of Brooke's historiographical work in T. Dixon, G. N. Cantor, and S. Pumfrey (2010) (eds) *Science and Religion: New Historical Perspectives* (Cambridge: Cambridge University Press), esp. pp. 1–19.
90. See F. M. Turner (1993) *Contesting Cultural Authority: Essays in Victorian Intellectual Life* (Cambridge: Cambridge University Press). For a slight revision of Turner's thesis, which figures the battle to have taken place on the grounds of science itself rather than 'science versus religion', see B. V. Lightman (2014) 'Science at the Metaphysical Society: Defining Knowledge in the 1870s', in B. V. Lightman and M. S. Reidy (eds) *The Age of Scientific Naturalism: Tyndall and His Contemporaries* (London: Pickering & Chatto), pp. 187–206.
91. For a discussion of natural theology that complicates this picture, see J. R. Topham (2004) 'Science, Natural Theology, and the Practice of Christian Piety in Early Nineteenth-Century Religious Magazines', in G. N. Cantor and S. Shuttleworth (eds) *Science Serialized: Representation of the Sciences in Nineteenth-Century Periodicals* (Cambridge, MA: MIT Press), pp. 37–66.
92. D. M. Knight, *Humphry Davy*, p. 52. The Tepidarians were a group of 'twenty-five of the most violent republicans of the day, who met to drink tea at Old Slaughter's Coffee House in St. Martin's Lane'. R. Hudson (1854) *Coleridge among the Lakes and Mountains*, p. 119, qtd. in D. M. Knight, *Humphry Davy*, p. 45.

93. D. M. Knight, *Humphry Davy*, pp. 42–56, at 45.
94. D. M. Knight, *Humphry Davy*, p. 35, emphasis added. Knight's interpretation of the *desideratum* to renounce the political associations of scientific materialism in the period are rather Whiggish, apparently suggesting that science necessarily progressed from the 'bad old days' of ideology, to its supposed 'value-neutrality' of modernity. However, as we can see in the case of Davy, observation was value-laden and supported particular social interests.
95. R. Carlile, *An Address*, p. 3.
96. J. F. W. Herschel (1987) *A Preliminary Discourse on the Study of Natural Philosophy* (Chicago: University of Chicago Press), p. 7.
97. W. Paley, M. Eddy, and D. M. Knight (2006) *Natural Theology: Or, Evidence of the Existence and Attributes of the Deity, Collected from the Appearances of Nature* (Oxford: Oxford University Press). While the importance of natural theology for the period is generally acknowledged, historians of science have mostly rejected the notions of R. M. Young (1985) *Darwin's Metaphor: Nature's Place in Victorian Culture* (Cambridge: Cambridge University Press) and S. F. Cannon (1978) *Science in Culture: The Early Victorian Period* (Kent, UK: Dawson) of a relatively homogeneous natural theology as the exclusive basis for natural philosophy. In particular, in J. R. Topham (1998) 'Beyond the "Common Context": The Production and Reading of the *Bridgewater Treatises*', *Isis* 89, 233–62, at pp. 233–37, Topham argues that R. M. Young was mistaken to the degree that he ignored both the heterogeneity within highbrow culture, and excluded the meanings imputed to such texts as the *Bridgewater Treatises* by cross-cultural (especially working-class and artisan) readers.
98. J. H. Wiener, *Radicalism and Freethought*, pp. 113–114.
99. J. H. Wiener, *Radicalism and Freethought*, pp. 114, and 119, note 46.
100. *Republican* (1826) 14.18, 577.
101. Here I am following closely J. Epstein, *Radical Expression*, chapter on Carlile and the Zetetic movement.
102. J. Epstein, *Radical Expression*, pp. 123–36, esp. at 125.
103. *Republican*, qtd. in J. Epstein, *Radical Expression*, p. 130.
104. E. Royle (1974) *Victorian Infidels: The Origins of the British Secularist Movement, 1791–1866* (Manchester: University of Manchester Press; Totowa, NJ: Rowman & Littlefield).
105. J. Epstein, *Radical Expression*, p. 131.
106. J. Epstein, *Radical Expression*, p. 127.
107. R. Carlile, qtd. in J. Epstein, *Radical Expression*, p. 21.
108. It is possible to see the radicalism of Carlile and the useful knowledge movement as drawing from the same well of philosophical radicalism, rather than one stemming from the other. Robbins traces the intellectual legacy of what he variously terms the 'political economy party' educationists, the 'philosophical radicals', 'the nineteenth-century ideologues', or the educational 'democrats' to some of the same sources that Carlile drew on for his epistemology and educational politics. Robbins traces the intellectual and political connections of the 'democrat' educationists and reformers, such as Henry Brougham, James Mill (and later, John Stuart Mill), Jeremy Bentham, Francis Place, Thomas Pole, etc., to the eighteenth-century political economists and social environmentalists (see especially the introduction and chapter 1). He does an excellent job in tracing the

intellectual history of 'the popular enlightenment', which for him included the useful knowledge movement (especially in chapter 2). My study is less concerned with intellectual forebears and connections, as it is with *representation* of those themes, especially in periodical formats.

109. *Working Man's Friend and Family Instructor* (1850) 1, 35.
110. *Mechanic's Magazine* (1823) Vol. 1, 14.
111. G. Lukács (1971) *History and Class Consciousness: Studies in Marxist Dialectics* (London: Merlin Press), pp. 149–222. For feminist appropriations of standpoint epistemology, see S. Harding (1986) *The Science Question in Feminism* (Ithaca: Cornell University Press), pp. 136–62.
112. V. H. B. Brougham (1825) *Practical Observations upon the Education of the People: Addressed to the Working Classes and their Employers* (London: Printed by Richard Taylor and sold by Longman, Hurst, Rees, Orme, Brown, and Green, for the benefit of the London Mechanics Institution), at p. 10. Dr. George Birkbeck, whose lectures to mechanics and artisans at the Anderson Institution in Glasgow evolved into the first Mechanics Institute in Great Britain '*established by mechanics by mechanics themselves*' (*Mechanic's Magazine* (1823) Vol. 1, 101, emphasis in original; Hudson), was not so 'sanguine in [his] expectations, that by a course of instruction ... one artist will be directed to the discovery of anything which is essential or important in his particular department' ('The Philosophical and Chemical Lectures', qtd. in *Mechanic's Magazine* 1, 116).
113. *Mechanic's Magazine* (1823) Vol. 3, p. ii.
114. For treatments of the Rotunda as employed by Carlile, see J. Wiener, *Radicalism and Freethought*, pp. 165–72; I. D. McCalman (1992) 'Popular Irreligion in Early Victorian England: Infidel Preachers and Radical Theatricality in 1830s London', in R. W. Davis and R. J. Helmstadter (eds) *Religion and Irreligion in Victorian Society: Essays in Honor of R. K. Webb* (London and New York: Routledge), pp. 62–78; and C. Parolin (2010) *Radical Spaces: Venues of Popular Politics in London, 1790–1845* (Canberra, ACT: ANU E Press), pp. 213–72.
115. The term is from E. Troeltsch (1931) *The Social Teachings of the Christian Churches* (New York: MacMillan) and is used to describe a kind of non-affiliated individual religiosity.
116. D. Nash, in 'Reassessing the "Crisis of Faith"', p. 72.
117. B. V. Qualls, *Secular Pilgrims*, pp. 791–806, at 803.

2 *Principles of Geology*: A Secular Fissure in Scientific Knowledge

1. C. Taylor (2007) *A Secular Age* (Cambridge, MA: Belknap Press of Harvard University Press).
2. O. Chadwick (1975) *The Secularization of the European Mind in the Nineteenth Century* (Cambridge: Cambridge University Press), pp. 161–90.
3. See for example D. Nash (2004) 'Reconnecting Religion with Social and Cultural History: Secularization's Failure as a Master Narrative', *Cultural and Social History* 1, 302–25.
4. In particular, P. L. Berger, once an important secularization theorist, reversed his long-standing position on secularization in (1999) *The Desecularization of*

the World: Resurgent Religion and World Politics (Washington, DC: Ethics and Public Policy Center), esp. at pp. 1–18.

5. L. Schwartz (2013) *Infidel Feminism: Secularism, Religion and Women's Emancipation, England 1830–1914* (Manchester, UK: Manchester University Press), p. 19.
6. F. M. Turner (2010) 'The Late Victorian Conflict of Science and Religion as an Event in Nineteenth-Century Intellectual and Cultural History', in T. Dixon, G. Cantor, and S. Pumfrey (eds) *Science and Religion: New Historical Perspectives* (Cambridge: Cambridge University Press), p. 88.
7. T. Dixon (2010) similarly argues against the reification of the categories of 'science' and 'religion'. See T. Dixon (2010) 'Introduction', in *Science and Religion*, pp. 4–5.
8. T. Dixon, 'Introduction', p. 9.
9. C. Lyell (1990) *Principles of Geology, Being an Attempt to Explain the Former Changes of the Earth's Surface, by Reference to Causes Now in Operation*, M. J. S. Rudwick (ed.) 3 vols (1830–1833), facsimile of the first edition, (Chicago; London: University of Chicago Press), Vol. 1, p. 4. As I refer to two different editions of Lyell's *Principles*, I will include the year of publication to differentiate them.
10. C. Lyell (1997) *Principles of Geology* (abridged), J. A. Secord (ed.) (London: Penguin Books), p. 9; C. Lyell (1990) *Principles of Geology*, Vol. 1, pp. 23–24.
11. Lyell (1990) *Principles of Geology*, Vol. I, pp. 71–72.
12. Editorial insertion, Lyell (1997) *Principles of Geology*, p. 9. William Whewell introduced the term 'Uniformitarian' as the antipode of 'Catastrophism' in an 1832 review of Lyell's *Principles of Geology*. Historians of science have generally recognized three aspects of Lyell's Uniformitarianism. These include 'non-progressionism' (the rejection of narratives of progress in the organic or inorganic world), 'actualism' (the tenet that past changes in the earth's surface must be explained with reference to forces now acting), and 'gradualism' (the tenet that change happens over long periods of time and is not catastrophic). It is the last of these that has come to characterize the Victorian period, and to refer to the analogy between the methodology of change in science, literature, and politics. Thus, the binary has become 'catastrophism/gradualism', as opposed to the original pairing of Catastrophist/Uniformitarian. By 'Catastrophism', Whewell meant the view that the earth had been geologically altered by sudden, violent, and short-lived events. One such unrepeatable catastrophe that was often debated was the global flood known as the Mosaic flood or 'Noah's flood'. Uniformitarianism referred to Lyell's version of uniformity, which excluded all but existing causes for explaining past geological phenomena.
13. A. Johns (1998) *The Nature of the Book: Print and Knowledge in the Making* (Chicago: University of Chicago Press), p. 35.
14. S. J. Gould (1987) *Time's Arrow, Time's Cycle: Myth and Metaphor in the Discovery of Geological Time* (Cambridge, MA: Harvard University Press), passim. Lyell's version of the history of geology had been adopted by historians from the nineteenth century on, including W. Whewell (1836–7), A. Ramsay (1848), Sir A. Geikie (1897), T. G. Bonney (1895), Sir E. B. Bailey (1962), C. C. Gillespie (1951), T. Kuhn (1962), and L. G. Wilson (1972). See R. Porter (1976) 'Charles Lyell and the Principles of the History of Geology', *British*

Journal for the History of Science 9.2, 91–103; part 2, 91, 100, notes 2–6; and M. T. Greene (1982) *Geology in the Nineteenth Century: Changing Views of a Changing World* (Ithaca, NY: Cornell University Press).

15. See, for example, the chapter on Hutton and Werner in M. T. Greene, *Geology in the Nineteenth Century*, pp. 19–45.
16. See R. Porter (1976) 'Charles Lyell and the Principles of the History of Geology', *British Journal for the History of Science* 9.2, part 2, 91. Porter calls Lyell's history 'catastrophist' (p. 98). See also W. F. Cannon (later known as S. F. Cannon) (1976) 'Charles Lyell, Radical Actualism, and Theory', *British Journal for the History of Science* 9.2, 104–20, at 108. Cannon reiterates her earlier claim that Lyell's history is 'erroneous propaganda' and 'historical romance'. The July 1976 issue of *British Journal for the History of Science* (9.32, part 2) was the Lyell Centenary Issue, a record of the conference in honor of the centenary of Lyell's death, held at Imperial College, King's College, the Geological Society, and the Royal Institution, London, 1–5 September 1975.
17. The biography is L. G. Wilson (1972) *Charles Lyell: The Years to 1841: The Revolution in Geology* (New Haven: Yale University Press). See esp. pp. 278–93 (for Vol. 1) and 328–39 (for Vol. 2). The scholarly introductions to the *Principles* are by M. J. S. Rudwick (1990) in *Principles of Geology*, pp. vii–lviii; and J. A. Secord (1997) *Principles of Geology*, pp. ix–xliii.
18. S. Sheets-Pyenson, qtd. in W. H. Brock and A. J. Meadows (1984) *The Lamp of Learning: Taylor & Francis and the Development of Science Publishing* (London, Philadelphia: Taylor & Francis), pp. xi–xii.
19. A. Rauch (2001) *Useful Knowledge: The Victorians, Morality and the March of the Intellect* (Durham, NC: Duke University Press), *passim*.
20. J. A. Secord (1997) 'Introduction', *Principles of Geology*, p. xv, emphasis added. As R. Cooter and S. Pumfrey (1994) have noted in 'Separate Spheres and Public Places: Reflections on the History of Science Popularisation and Science in Popular Culture', *History of Science* 32, 237–67, the term 'popular' can signify both 'non-elite' and 'common to all levels of society' (p. 252). I am using the term in the less specific sense, and mean by use of the term 'popular' to emphasize the importance of *participation* in knowledge entertained by this widened circle of enthusiasts.
21. J. Klancher (2013) *Transfiguring the Arts and Sciences: Knowledge and Cultural Institutions in the Romantic Age* (Cambridge, UK; New York: Cambridge University Press), pp. 127–28.
22. Cooter and Pumfrey note that elite culture is rarely, if ever, homogenous, and the natural sciences may represent both a threat to and/or a confirmation of established values and beliefs. See R. Cooter and S. Pumfrey, 'Separate Spheres', 237–67, esp. at 251.
23. [C. Lyell] (1825–26) 'The London University', *Quarterly Review* 33, 257–75; [C. Lyell] (1826) 'Scientific Institutions', *Quarterly Review* 34, 153–79; [C. Lyell] (1827) 'State of the Universities', *Quarterly Review* 36, 216–68.
24. C. Lyell, 'The London University', 257–75, at 259.
25. [H. Brougham] (1825) 'New University in London', *The Edinburgh Review, or Critical Journal* 42, 346–67, esp. at 354–55; [H. Brougham] (1825–26) 'The London University', *The Edinburgh Review, or Critical Journal* 43, 315–41.
26. C. Lyell to W. Whewell (20 February 1831) qtd. in L. G. Wilson, *Charles Lyell*, p. 308.

27. J. A. Secord (1997) 'Introduction', *Principles of Geology*, p. xiii. Openly advocating political reform would have led to his estrangement from his father, on whom he still relied for emotional and financial support. His career as a barrister did not earn him much money. See L. G. Wilson, *Charles Lyell*, esp. pp. 135–82.
28. C. Lyell, 'Scientific Institutions', 153–79; C. Lyell, 'State of the Universities', 216–68.
29. While historians of science have generally seen the nineteenth-century development of science as a second scientific revolution, the first occurring in the seventeenth century, later commentators have suggested that it was in fact the first scientific revolution. See A. Cunningham and N. Jardine (eds) (1990) 'Introduction: The age of reflexion', in *Romanticism and the Sciences* (Cambridge, UK; New York: Cambridge University Press), pp. 1–9; and D. Knight (1992) *Humphry Davy: Science & Power* (Oxford, UK; Cambridge, USA: Blackwell), p. xii.
30. William Whewell later made the same connection; see W. Whewell (1833) *Astronomy and General Physics Considered with Reference to Natural Theology*, Third Bridgewater Treatise (London: W. Pickering). Whewell connected commerce and colonialism to the growth of knowledge. The stimulus to trade and colonialism was the variety of vegetables and animals that God had placed in the various regions of the earth. In fact, God had placed them as such in order to encourage such trade and colonialism: 'The intercourse of nations in the way of discovery, colonization, commerce; the study of the natural history, manners, institutions of foreign countries; lead to most numerous and important results. Without dwelling upon this subject, it will probably be allowed that such intercourse has a great influence upon the comforts, the prosperity, the arts, the literature, the power, of the nations which thus communicate. Now the variety of the productions of different lands supplies both the stimulus to this intercourse, and the instruments by which it produces its effects. The desire to possess the objects or the knowledge which foreign countries alone can supply, urges the trader, the traveler, the discoverer to compass land and sea; and the advantages of civilization consists almost entirely in the cultivation, the use, the improvement of that which has been received from other countries'. (See pp. 62–63).
31. C. Lyell, 'Scientific Institutions', 155.
32. C. Lyell, 'Scientific Institutions', 159.
33. C. Lyell, 'Scientific Institutions', 162–63.
34. H. Brougham, 'New University in London', 346–67, esp. at 354–55.
35. C. Lyell, 'Scientific Institutions', 163–71.
36. C. Lyell, 'Scientific Institutions', 172, emphasis mine.
37. C. Lyell, 'Scientific Institutions', 179. In dwelling on the importance of the provinces and colonies, Lyell advocated a system that mirrored what became the gentleman's imperial relationship with his sources – a relationship marked by the appropriation of often working-class, provincial and colonial knowledge, by metropolitan elite science. For an analysis of such relationships between gentleman naturalists and their mostly artisan and rural correspondents, see A. Secord (1994) 'Corresponding Interests: Artisans and Gentlemen in Nineteenth-Century Natural History', *British Journal for the History of Science* 27, 383–408.

38. In a recent essay, B. V. Lightman concludes that the relationship between the scientific naturalists and religionists was more complicated than F. M. Turner suggested in his groundbreaking monograph, *Contesting Cultural Authority* (1993). Lightman suggests that neither the theists nor the scientific naturalists viewed the struggle in terms of 'science versus religion'. Rather, the theists fought on the grounds of science itself, unwilling to yield scientific authority to the scientific naturalists. For their part, the scientific naturalists argued only against *theology*'s truth claims, not against *religion*'s role in the private realm of feeling. See B. V. Lightman (2014) 'Science at the Metaphysical Society: Defining Knowledge in the 1870s', in B. V. Lightman and M. S. Reidy (eds) *The Age of Scientific Naturalism: Tyndall and His Contemporaries* (London; Brookfield, Vermont: Pickering & Chatto), pp. 187–206.
39. C. Lyell to G. Mantell (22 June 1826) [Mrs.] K. M. Lyell (ed.) (1881) *Life Letters and Journals of Charles Lyell* (Farnborough, Hants: Gregg), Vol. 1, pp. 164–65.
40. J. A. Secord (1997) 'Introduction', *Principles of Geology*, p. xiii.
41. C. Lyell to J. Murray, 2 May 1827, qtd. in L. G. Wilson, *Charles Lyell*, p. 164, emphasis in original.
42. C. Lyell to J. Murray, 2 May 1827, qtd. in L. G. Wilson, *Charles Lyell*, p. 164.
43. C. Lyell to J. Murray, 2 May 1827, qtd. in L. G. Wilson, *Charles Lyell*, p. 164.
44. C. Lyell, 'State of the Universities', p. 218.
45. C. Lyell, 'State of the Universities', p. 218.
46. C. Lyell, 'State of the Universities', p. 220.
47. C. Lyell, 'State of the Universities', p. 221.
48. See, for example, the vitalism-materialism debate: Anon. (July 1819) 'Abernethy, Lawrence, &c. on the Theories of Life', *Quarterly Review* 22, 1–34.
49. 'John Barrow, traveler and author was ... a frequent contributor to the *Quarterly Review*' (L. G. Wilson, *Charles Lyell*, p. 159, note 55).
50. C. Lyell to his sister, Caroline Lyell, qtd. in L. G. Wilson, *Charles Lyell*, p. 159.
51. L. G. Wilson, *Charles Lyell*, p. 270.
52. L. G. Wilson, *Charles Lyell*, pp. 273–74; 327. The reviewers were George Scrope and William Whewell.
53. S. Sheets-Pyenson (1976) 'Low Scientific Culture in London and Paris, 1820–1875', unpublished doctoral dissertation, University of Pennsylvania, p. 47; A. Rauch, *Useful Knowledge*, p. 43.
54. The Murray publishing firm was famous in the nineteenth century and has attracted the attention of scholars ever since. By the 1820s, John Murray II was the doyen of English publishers. The firm boasted a list of authors that included Jane Austen, John Wilson Croker, Leigh Hunt, Samuel Taylor Coleridge, Walter Scott, Madame de Stael, Thomas Moore, Sir John Franklin, Washington Irving, and Thomas Robert Malthus, among others. Murray's reputation amongst authors was formidable. His courage in literary speculation earned the name given him by Lord Byron of 'the Anak [king] of stationers'. His acknowledged authority in the industry prompted Lyell to refer to him as 'absolute John'. For a history and correspondence of the Murray publishing house, see S. Smiles (1891) *A Publisher and His Friends: Memoir and Correspondence of the Late John Murray, with an Account of the Origin and Progress of the House, 1768–1843*, 2 vols (London: J. Murray).
55. C. Knight (1854) *The Old Printer and the Modern Press* (London: J. Murray), p. 243.

56. S. Smiles, *A Publisher and His Friends*, Vol. 1, pp. 295–96.
57. J. A. Secord (2000) *Victorian Sensation: The Extraordinary Publication, Reception, and Secret Authorship of Vestiges of the Natural History of Creation* (Chicago: University of Chicago Press), pp. 48–49.
58. S. Bennett (1976) 'John Murray's Family Library and the Cheapening of Books in Early Nineteenth Century Britain', *Studies in Bibliography* 29, 140–67, esp. at 141; J. A. Secord, *Victorian Sensation*, p. 50.
59. S. Bennett, 'John Murray's Family Library', p. 140.
60. S. Bennett, 'John Murray's Family Library', p. 141.
61. S. Bennett, 'John Murray's Family Library', p. 141; S. Smiles, *A Publisher and His Friends*, Vol. II, p. 296.
62. Bennett, 'John Murray's Family Library', pp. 141–42.
63. J. A. Secord (1997) 'Introduction', *Principles of Geology*, p. xiv.
64. The first edition sold for about twenty shillings per volume and included a few expensive copper engravings. Its use of the much cheaper wood engravings marked an early instance of what would become a rising trend.
65. J. A. Secord (1997) 'Introduction', *Principles of Geology*, p. xiv. The Family Library issues were priced at five shillings each.
66. M. J. S. Rudwick (1990) 'Introduction', *Principles of Geology*, pp. xi–xii; J. A. Secord (1997) 'Introduction', *Principles of Geology*, p. xiv. However, As M. J. S. Rudwick (1985) points out, especially as they were reprinted by the weekly *Athenaeum* and the commercial *Philosophical Magazine*, and in foreign journals, the *Proceedings* were more likely than the *Transactions* to be read by a wider audience. See M. J. S. Rudwick (1985) *The Devonian Controversy: The Shaping of Scientific Knowledge Among Gentlemanly Specialists* (Chicago: University of Chicago Press), p. 26.
67. C. Lyell to M. Horner (17 February 1832) qtd. in L. G. Wilson, *Charles Lyell*, p. 344.
68. J. A. Secord, *Victorian Sensation*, pp. 55–56.
69. C. Lyell to his father (10 April 1827), in [Mrs.] K. M. Lyell, *Life, Letters and Journals of Charles Lyell*, Vol. I, pp. 169–71, at 170; Mrs. Marcet (1809) *Conversations on Chemistry, in Which the Elements of that Science Are Familiarly Explained, and Illustrated by Experiments and Plates: To Which Are Added, Some Late Discoveries on the Subject of the Fixed Alkalies* (New Haven, CT: From Sidney's Press, for Increase Cooke & Co.).
70. D. Knight, *Humphry Davy*, p. 12.
71. [G. Penn] (1828) *Conversations on Geology: Comprising a Familiar Explanation of the Huttonian and Wernian Systems: The Mosaic Geology* (London: Printed for S. Maunder).
72. M. J. S. Rudwick (1990) 'Introduction', *Principles of Geology*, p. xi, note 3.
73. J. A. Secord, *Victorian Sensation*, pp. 42–52, esp. at 42.
74. J. F. W. Herschel (1830) *A Preliminary Discourse on the Study of Natural Philosophy* (Chicago: University of Chicago Press), p. vi; M. Ruse (1976) 'Charles Lyell and the Philosophers of Science', *British Journal for the History of Science* 9, 121.
75. H. Davy, Sir, and J. Davy (1830) *Consolations in Travel, or, the Last Days of a Philosopher* (London: J. Murray).
76. T. Hope (1831) *Essay on the Origin and Prospects of Man* (London: Murray).
77. M. Bartholomew (1973) 'Lyell and Evolution: An Account of Lyell's Response to the Prospect of an Evolutionary Ancestry for Man', *British Journal for the*

History of Science 6, 261–303; M. Bartholomew (1976) 'The Non-Progression of Non-Progression: Two Responses to Lyell's Doctrine', *British Journal for the History of Science* 9, 166–74; P. Corsi (1978) 'The Importance of French Transmutationist Ideas for the Second Volume of Lyell's *Principles of Geology*', *British Journal for the History of Science* 11, 221–44.

78. A. Desmond (1987) 'Artisan Resistance and Evolution in Britain, 1819–1848', *Osiris* 3, 2nd Series, 77–110.
79. A. Desmond (1989) *The Politics of Evolution: Morphology, Medicine, and Reform in Radical London* (Chicago: University of Chicago Press).
80. J. A. Secord, *Victorian Sensations*, p. 43, discusses the importance of familiar authors for selling works of knowledge.
81. C. Lyell (1990) *Principles*, Vol. 1, Chapters 7 and 8; and Chapter 9, pp. 104–66.
82. According to Secord, Lyell read passages before the book went to press: 'Lyell must have been shown the relevant passages in proof and realized that views similar to his own were suspected of tending to materialism': See J. A. Secord (1997) 'Introduction', *Principles of Geology*, p. xxxi. Lyell's biographer has him purchasing the book and reading it in early March; see L. G. Wilson, *Charles Lyell*, p. 284.
83. C. Lyell (1990) *Principles of Geology*, Vol. 1, pp. 144–45. See the note following the quotation of Davy by Lyell, below.
84. Lyell's penchant for misquoting and failing to give due credit to sources is legend, and criticisms began soon after publication. See Rev. W. D. Conybeare (1830) 'On Mr. Lyell's "Principles of Geology"', *The Philosophical Magazine and Annals of Philosophy* n.s. 8, 215–19, where Conybeare accuses Lyell of appropriating his classical allusions without proper citation.
85. H. Davy, *Consolations in Travel*, p. iii.
86. H. Davy, *Consolations in Travel*, pp. 27–28.
87. H. Davy, *Consolations in Travel*, p. 42.
88. H. Davy, *Consolations in Travel*, p. 40.
89. H. Davy, *Consolations in Travel*, p. 57.
90. H. Davy, *Consolations in Travel*, pp. 56–58. W. Whewell, *Astronomy and General Physics Considered with Reference to Natural Theology*.
91. H. Davy, *Consolations in Travel*, pp. 58–60. The American editor of *Consolations* objected to the exclusive survival of the intellect as opposed to the 'spiritual' in Davy's book (p. 71). Throughout the text, the American editor corrects the theology of Davy's 'ideal' characters, and especially the heterodoxies contained in the vision. As the 'Note to the American Edition' states, 'Most of the notes have been added, with the view of correcting some theological errors which, had the distinguished author lived to revise his work, it is believed he would have modified or erased himself' (iv). Apparently this editor missed the point that Davy had elaborated these positions in order to more thoroughly refute them later.
92. H. Davy, *Consolations in Travel*, pp. 59–60, emphasis in original.
93. H. Davy, *Consolations in Travel*, p. 73.
94. H. Davy, *Consolations in Travel*, pp. 132, 142.
95. L. G. Wilson, *Charles Lyell*, p. 285, note 3.
96. H. Davy, *Consolations in Travel*, p. 147, emphasis added.

97. H. Davy, *Consolations in Travel*, p. 148, emphasis mine. Attributing fossils to a 'plastic force' thought to create them from stones was ridiculed by C. Lyell (1990) *Principles of Geology*, Vol. 1, p. 23. The 'theory' that God created and buried such fossils to foil geological speculators had been proposed by some theologians.
98. L. G. Wilson, *Charles Lyell*, pp. 284–93; C. Lyell (1990) *Principles*, Vol. 1, pp. 104–43.
99. C. Lyell (1990) *Principles of Geology*, Vol. 1, p. 144.
100. Qtd. in C. Lyell (1990) *Principles of Geology*, Vol. 1, pp. 144–45.
101. C. Lyell (1990) *Principles of Geology*, Vol. 1, p. 145, emphasis mine. The opening sentence of Lyell's quotation is actually from page 147 of *Consolations*, a passage from later in the published text than all other parts. The clause beginning with 'in those strata' is from page 143. The clause 'and whoever dwells upon this subject', is from page 145, and actually begins a new sentence there. 'In the oldest secondary strata' is found on page 147, and ends with 'the existence of man', where Lyell begins his quotation: 'You must allow that it is impossible to defend the proposition ...'. As I have suggested, Lyell misquoted, or else he was looking at manuscript or a publisher's proof. The following discussion argues that Lyell misquoted Davy for argumentative effect.
102. C. Lyell (1990) *Principles of Geology*, Vol. 1, p. 144. Lyell interchanged the term 'progressive development' with 'successive development' throughout chapter nine, although the terms surely suggest distinct meanings. 'Progressive development' might imply transmutation while 'successive development' could be limited to suggest serial, special creations.
103. C. Lyell (1990) *Principles of Geology*, Vol. 1, p. 147.
104. C. Lyell (1990) *Principles of Geology*, Vol. 1, pp. 148–50.
105. J. A. Secord (1997) 'Introduction', *Principles of Geology*, p. xxxi. T. Hope, *An Essay on the Origin and Prospects of Man*, 3 vols.
106. J. A. Secord (1997) 'Introduction', *Principles of Geology*, p. xxxi.
107. F. M. Turner (1993) *Contesting Cultural Authority: Essays in Victorian Intellectual Life* (Cambridge; New York, NY: Cambridge University Press); B. V. Lightman and M. S. Reidy, (eds) *The Age of Scientific Naturalism: Tyndall and His Contemporaries*.

3 Holyoake and Secularism: The Emergence of 'Positive' Freethought

1. E. Royle (1974) *Victorian Infidels: The Origins of the British Secularist Movement, 1791–1866* (Manchester: University of Manchester Press); and E. Royle (1980) *Radicals, Secularists, and Republicans: Popular Freethought in Britain, 1866–1915* (Manchester: University of Manchester Press).
2. Recent studies on Secularism that have at least addressed the relationship of Secularism to the secular, secularization and secularism considered more broadly include L. Schwartz (2013) *Infidel Feminism: Secularism, Religion and Women's Emancipation, England 1830–1914* (Manchester, UK: Manchester University Press), pp. 18–22; and M. Rectenwald (2013) 'Secularism and the

Cultures of Nineteenth-Century Scientific Naturalism', *British Journal for the History of Science* 46.2, 231–54, at 232.
3. E. Royle (1980) *Radicals, Secularists, and Republicans*, pp. xi–xii.
4. O. Chadwick (1975) *The Secularization of the European Mind in the Nineteenth Century* (Cambridge, UK: Cambridge University Press), pp. 88–106.
5. C. Taylor (2007) *A Secular Age* (Cambridge, MA: Belknap Press of Harvard University Press).
6. G. Smith (2008) *A Short History of Secularism* (London: I. B. Tauris), pp. 174–75.
7. The term 'freethought' might occasion some confusion for contemporary readers, especially in connection with a history of a discourse in which 'free will' was considered a meaningless term (as in Robert Owen's necessarianism). Freethought did not mean to suggest unfettered, willy-nilly thought free of environmental determination. J. M. Robertson captured something of the meaning for the period when he described it in the Preface to his history of the movement: '"Free" is a term of relation, of antithesis. It is significant only in mental relation to "unfree," and it takes root and application in regard to human *action*, which may be either free or unfree, in respect of either moral or physical coercion ... [t]he term "freethought" is found to be a serviceable label, connoting as it does the perpetual process of conflict between sacrosanct and critical opinion'. See J. M. Robertson (1930) *A History of Freethought in the Nineteenth Century* (New York: G. P. Putnam's Sons), pp. xxxiv–xxxv, emphasis in original.
8. Secularism will be defined and discussed throughout this chapter, drawing from Holyoake's own words on the subject, especially George Jacob Holyoake (1871) *The Principles of Secularism Illustrated* (London: Austin & Co.).
9. L. Grugel (1976) *George Jacob Holyoake: A Study in the Evolution of a Victorian Radical* (Philadelphia: Porcupine Press), pp. 2–3. In addition to Grugel's biography, for biographical sketches of Holyoake, see E. Royle (1974) *Victorian Infidels*, esp. at pp. 3–6, 72–74, and 312; J. McCabe (1908) *Life and Letters of George Jacob Holyoake*, 2 vols (London: Watts & Co.); and J. M. Wheeler (1889) *A Biographical Dictionary of Freethinkers of All Ages and Nations* (London: Progressive Pub. Co.).
10. G. J. Holyoake (1896) *English Secularism: A Confession of Belief* (Chicago: Open Court Pub. Co.), pp. 45–49.
11. According to the OED, the word 'secular' had referred to worldly as opposed to spiritual concerns since as early as the late thirteenth century. The word was part of theological discourse. The first usage applied to clergy who lived outside of monastic seclusion. But never before Holyoake's mobilization had it been used as an adjective to describe a set of principles or as a noun to positively delineate principles of morality or epistemology.
12. Holyoake consistently differentiated Secularism from secular education, for example, although arguably his notion of Secularism owed something to the secular education movement underway at this time.
13. In C. Campbell (1971) *Toward a Sociology of Irreligion* (London: Macmillan), p. 54, Campbell referred to these two approaches as the 'substitutionist' (Holyoake) and 'eliminationist' (Bradlaugh) camps.
14. B. Lightman (1989) 'Ideology, Evolution and Late-Victorian Agnostic Popularizers', in J. Moore (ed.) *History, Humanity and Evolution: Essays for John C. Greene* (Cambridge: Cambridge University Press), pp. 285–309, at 287–88.

15. E. Royle, *Victorian Infidels*, p. 76, argued that 'the *Oracle* reads like the *Republican* of 1822, and, like Carlile, the editors of the *Oracle* were eventually to calm down. But also like him, they were first to suffer imprisonment.' In J. Secord (2000) *Victorian Sensation: The Extraordinary Publication, Reception, and Secret Authorship of Vestiges of the Natural History of Creation* (Chicago: University of Chicago Press), p. 303, note 10, Secord notes the relative scholarly neglect of the radical periodicals of the 1840s and 50s. For a discussion of the print culture of 1840s infidelity and early Secularism, see M. Rectenwald, 'Secularism and the Cultures of Nineteenth-Century Scientific Naturalism', pp. 235–38.
16. J. M. Robertson, *A History of Freethought in the Nineteenth Century*, p. xxviii.
17. P. Hollis (1970) *The Pauper Press: A Study in Working-Class Radicalism of the 1830s* (Oxford: Oxford University Press), p. 153.
18. Neville Wood to Richard Carlile (21 June 1839) quoted in P. Hollis, *The Pauper Press*, p. 155.
19. A. Desmond (1987) 'Artisan Resistance and Evolution in Britain, 1819–1848', *Osiris* 3, 2nd Series (Chicago: The University of Chicago Press), pp. 77–110, at 85. S. Budd (1977) *Varieties of Unbelief: Atheists and Agnostics in English Society, 1850–1960* (London: Heinemann Educational Books), p. 10, similarly notes the failures of Chartism, Trade Unionism and Owenism as contributory factors for the support of Secularism during the later 1840s.
20. E. Royle, *Victorian Infidels*; A. Desmond, 'Artisan Resistance', p. 85; and S. Budd, *Varieties of Unbelief*, pp. 29–30. The major cause of ire amongst radical atheists was the attempt by the Owenite Central Committee to register the socialists as a Protestant sect called 'Rational Religionists' in a move designed to outmaneuver Anglicans who attempted to legally block the socialists' Sunday activities. 'The Religion of Truth', or 'Rational Religion' became the new name for the system for the scientific formation of human character promoted by Robert Owen; see R. Owen (1970) 'The Formation of Character', in *The Book of the New Moral World in Seven Parts [1842]* (New York: Augustus M. Kelley), Second Part, p. 38. The radical infidels considered this declaration of religion to be a breach of socialist principles; see for example, *The Oracle of Reason: or, Philosophy Vindicated* (1842) 1, 57–59, 73–75, 81–82, 89–91.
21. Anglican ministers, led by Henry Phillpotts, the Bishop of Exeter, pressured the House of Lords to abolish the Owenite practice of Sunday meetings in the Halls of Science, where collections were taken. Rather than risking legal action, the Central Board required its social missionaries to swear the oath to the Crown required of all Nonconformist ministers. Many Social Missionaries, such as Lloyd Jones, did not see any contradiction in the notion of Owenism as a kind of Protestant sect.
22. *Oracle* (1842) 1, ii. In G. J. Holyoake (1892) *Sixty Years of an Agitator's Life*, 2 vols (London: T. F. Unwin) Vol. 1, p. 142, Holyoake described Chilton as 'a cogent, solid writer, ready for any risk, and the only absolute atheist I have ever known'. Presumably, Carlile's dalliances with deism allowed the *Oracle* to claim to be the first exclusively atheist periodical to have appeared. Royle contradicts this in *Victorian Infidels*, but Robertson confirms it in *A History of Freethought in the Nineteenth Century*, pp. 73–75, claiming that 'Richard Carlile had always been a deist'.

23. The differences were many, such as the fact that Southwell was an artisan-class radical, not a university-educated philosopher trained in German philosophy. But J. M. Robertson, *A History of Freethought in the Nineteenth Century*, p. 75, compares the atheism in the *Oracle* to positions developed by Feuerbach. For biographical sketches of Southwell, see A. Desmond 'Artisan Resistance'; E. Royle, *Victorian Infidels*, pp. 69–73; and J. M. Robertson, *A History of Freethought in the Nineteenth Century*, Vol. 1, p. 73.
24. K. Marx (1998) *The German Ideology: Including Thesis on Feuerbach* (Amherst, NY: Prometheus Books), pp. 29–105, esp. at p. 65.
25. K. Marx, *The German Ideology*, p. 36.
26. C. Southwell (1841) *Oracle* 1, 1.
27. C. Southwell (1841) *Oracle* 1, 28, emphasis in original.
28. See Oracle (1842) 1, 2–4, 19–21, 27–29, 35–37. As C. Southwell and W. Carpenter noted in (1842) *The Trial of Charles Southwell: (Editor of "the Oracle of Reason") for Blasphemy, Before Sir Charles Wetherall [i.e. Wetherell] Recorder of the City of Bristol, January the 14th, 1842* (London: Hetherington), pp. 2–7, several of these articles ('Is There A God?') were also cited in the indictment as counts of blasphemy.
29. C. Southwell (1841) *Oracle* 1, 25.
30. J. A. Secord, *Victorian Sensation*, p. 307.
31. C. Southwell (circa 1850) *Confessions of a Freethinker* (London), p. 66.
32. He remained there for seventeen days until an offer of bail was finally accepted. As Maltus Ryall pointed out in the Preface to C. Southwell and W. Carpenter, *The Trial of Charles Southwell*, p. iii, the offers of bail by several parties were rejected three times before the magistrates accepted 'the *same bail* which they had refused ten days previously'. The indictment stated that he had 'unlawfully and wickedly composed, printed and published a certain scandalous, blasphemous, and profane libel, of and concerning the Holy Scriptures and the Christian Religion'. The quote is from a handbill, which included the indictment, printed and distributed by his partners in Bristol. The reproduction of the indictment was illegal as well.
33. See in particular, article by 'Publicola' (19 December, 23 January, and 30 January 1842) *Weekly Dispatch*.
34. C. Southwell and W. Carpenter, *The Trial of Charles Southwell*, p. 102. Holyoake was the next to be imprisoned for his discussion after a Socialist lecture in Cheltenham. The third editor, Thomas Paterson, was next imprisoned for selling blasphemous placards in London. A fourth editor, George Adams, was next imprisoned for selling a copy of the *Oracle*. William Chilton then succeeded Adams, until the *Oracle* was superseded by the *Movement*, edited by Holyoake. See J. M. Robertson, *A History of Freethought in the Nineteenth Century*, Vol. 1, p. 73.
35. C. Southwell (1842) *Oracle* 1, 1.
36. M. Q. Ryall (1842) *Oracle* 1, 67. In O. Chadwick, *The Secularization of the European Mind*, p. 90, Chadwick also notes a 'shift to moderation' in freethought with Holyoake, as does S. Budd, *Varieties of Unbelief*, pp. 26–34. Although two additional installments of 'The Jew Book' were published and the term 'Jew Book' continued to be used in the *Oracle*, the 'manner' of the articles changed from outright denunciation of the Bible to more subtle criticism along with an admission that some parts of the Bible were morally useful. See (1842) *Oracle* 1, 289, 305, and 383.

37. O. Chadwick, *The Secularization of the European Mind*, p. 48. E. Gaskell (1848) *Mary Barton: A Tale of Manchester Life* (London: Chapman and Hall), chapter 37.
38. For Holyoake's moral conversion, see E. Royle, *Victorian Infidels*, pp. 76–77.
39. G. J. Holyoake (1842) *The Spirit of Bonner in the Disciples of Jesus, or, the Cruelty and Intolerance of Christianity Displayed in the Prosecution, for Blasphemy, of Charles Southwell, Editor of the Oracle of Reason: A Lecture* (London: Hetherington and Cleave), p. 4.
40. G. J. Holyoake (1842) *Oracle* 1, 81.
41. Holyoake's conversion seems to contradict the 'existential security hypothesis' of secularization. For an extended discussion of this hypothesis, see P. Norris and R. Inglehart (2011) *Sacred and Secular: Religion and Politics Worldwide* (West Nyack, NY: Cambridge University Press). Norris and Inglehart advance the 'existential security hypothesis' as the explanation for secularization or the lack thereof. According to this thesis, as populations become relatively secure economically and otherwise, religiosity tends to decline.
42. G. J. Holyoake (1842) *Oracle* 1, 81.
43. G. J. Holyoake (1842) *Oracle* 1, 81–82.
44. G. J. Holyoake (1842) *Oracle* 1, 81.
45. G. J. Holyoake (1842) *Oracle* 1, 81, emphasis in original.
46. Thomas Carlyle coined the phrase in chapter one of *Chartism* (1839). Friedrich Engels contributed to the discourse in 1844 with *The Condition of the Working Class in England*. See M. Levin (1998) *The Condition of England Question: Carlyle, Mill, Engels* (New York: St. Martin's Press).
47. P. Hollis, *The Pauper Press*, pp. 203–58, esp. at p. 204.
48. M. Q. Ryall (1843) 'The Belly and the Back', *The Movement and Anti-Persecution Gazette And Register of Progress: A Weekly Journal of Republican Politics, Anti-Theology and Utilitarian Morals* 1, 3–4.
49. The Anti-Persecution Union was formed primarily in response to the imprisonment for blasphemous libel of Charles Southwell and grew out of the 'Committee for the Protection of Mr. Southwell'. Subscriptions for the Union and its establishment were announced in *Oracle* (1842) 1, 72. Ryall was its first secretary; Holyoake became its secretary by 1843; see *Movement* (1843) 1, 5–7.
50. G. J. Holyoake, *Englishc Secularism*, p. 34.
51. Quoted in G. J. Holyoake, *English Secularism*, p. 34. See also G. J. Holyoake and C. Bradlaugh (1870) *Secularism, Scepticism, and Atheism: Verbatim Report of the Proceedings of a Two Nights' Public Debate between Messrs. G. J. Holyoake & C. Bradlaugh: Held at the New Hall of Science ... London, on the Evenings of March 10 and 11, 1870* (London: Austin), p. iv, where Holyoake traces the phrase to Madame de Staël, who attributed it in her 1818 *Considerations of the French Revolution* to M. Necker. Necker reputedly argued to the Constituent Assembly for the adoption of the English Constitution saying, 'Nothing is effectually destroyed for which we do not provide a substitute'. Holyoake suggests that Napoleon appropriated the phrase and converted it to the statement, '[n]othing is destroyed until it is replaced'.
52. The motto was carried as the epigraph following the title of each number.
53. G. J. Holyoake (1843) 'Prospectus of the *Movement*', qtd. in G. J. Holyoake, *English Secularism*, p. 46. In *English Secularism*, Holyoake identified this statement as the first indication of Secularism's 'rise and development'.

54. (1843) 'The Principles of Naturalism', *Movement* 1, 2.
55. (1843) 'The Principles of Naturalism', *Movement* 1, 2.
56. (1843) 'The Principles of Naturalism', *Movement* 1, 2.
57. G. J. Holyoake (1843) 'The Mountain Sermon', *Movement* 1, 42–43.
58. G. J. Holyoake, *Sixty Years of an Agitator's Life*, Vol. 1, pp. 204–08.
59. For accounts of Holyoake's trial, see G. J. Holyoake (1850) *The History of the Last Trial by Jury for Atheism in England: A Fragment of Autobiography* (London: Watson), and 'The Trial', in G. J. Holyoake, *Sixty Years of an Agitator's Life*, Vol. 1, pp. 157–63.
60. In 1861, Holyoake joined Bradlaugh as co-editor of the *National Reformer*, but differences in personality and approach made the partnership a short-lived one, which lasted only a few months. Holyoake published *The Counsellor* from 1863 to 1864, and attempted to revive the *Reasoner* from 1871 to 1872. (S. Budd, *Varieties of Unbelief*, pp. 40–44 and p. 297.)
61. G. Holyoake (1847) 'Letter to Paul Rodgers', *Reasoner* 3, 485–92, at 488.
62. C. Southwell (1847) 'Letter from Mr. Southwell', *Reasoner* 3, 577–81, at 579.
63. W. Chilton (1847) 'Letter from Mr. Chilton', *Reasoner* 3, 607–10, at 607.
64. For conflicts with Southwell, see (1847) 'Letter from Mr. Southwell', *Reasoner* 8, 579–81; his conflicts with Bradlaugh are characterized by the 1870 debate reproduced in G. J. Holyoake and C. Bradlaugh, *Secularism, Scepticism, and Atheism*.
65. S. Budd, *Varieties of Unbelief*, p. 38, describes the earlier prose of freethought as 'vivid, concrete and pungent', which she argues gave way to a 'more prolix, abstract and long-winded' prose by the late 1840s.
66. *Reasoner* (1846) 1, 1.
67. *Reasoner* (1848) 5, 1–2. That the article was intended to be the basis of a later book was evident from the outset, with references to 'such a book' in the first article. *Public Speaking and Debate* was published in 1866.
68. *Reasoner* (1848) 5, 1–2, emphasis in original.
69. E. Royle, *Victorian Infidels*, p. 74.
70. S. Budd, *Varieties of Unbelief*, p. 38.
71. It is interesting to consider whether or not British freethought might have developed into a movement, which, at least in terms of a class character and relative autonomy, might have resembled Marxism, had Holyoake not sought to broaden the class base of his materialist movement.
72. Less adequate (and even more cynical than Royle) is the explanation that Secularists like Holyoake merely aped their social superiors out of false consciousness. For the charge of false consciousness and aping, see R. Billington (1968) 'Leicester Secular Society, 1852–1920: A Study in Radicalism and Respectability', unpublished dissertation, University of Leicester, 1968; cited in S. Budd, *Varieties of Unbelief*, p. 36.
73. *Reasoner* (1847) 3, 611.
74. *Reasoner* (1847) 3, 611.
75. *Reasoner* (1848) 5, 2.
76. *Reasoner* (1848) 5, 2.
77. *Reasoner* (1848) 5, 1.
78. S. Budd, *Varieties of Unbelief*.
79. G. J. Holyoake, *English Secularism*, p. 49.
80. *Reasoner* (1849) 7, 33–37; and *Reasoner* (1849) 7, 49–53.

81. G. H. Lewes to G. Holyoake (8 August 1849), The National Co-operative Archive, Manchester (subsequently NCA).
82. G. J. Holyoake, *The History of the Last Trial by Jury for Atheism.*
83. J. McCabe, *Life and Letters of George Jacob Holyoake*, Vol. 1, p. 145; E. Royle, *Victorian Infidels*, 154–55; B. J. Blaszak (1988) *George Jacob Holyoake (1817–1906) and the Development of the British Cooperative Movement* (Lewiston, NY: The Edwin Mellen Press), p. 17; and R. Ashton (2008) *142 Strand: A Radical Address in Victorian London* (London: Vintage), pp. 8–9.
84. As is well known, Lewes was meanwhile having an affair with Marian Evans (soon to adopt the penname of George Eliot; see Chapter 6).
85. E. Royle, *Victorian Infidels*, p. 154.
86. T. Hunt to G. Holyoake (18 December 1849), NCA.
87. T. Hunt to H. Travis (21 October 1850), Holyoake Papers, Bishopsgate Institute Library, London.
88. T. Hunt to G. Holyoake (13 September 1852), NCA.
89. T. Hunt to G. Holyoake (18 December 1849), NCA.
90. T. Hunt to George Holyoake (18 December 1849), NCA.
91. E. Royle, *Victorian Infidels*, p. 158.
92. F. W. Newman (1854) *Catholic Union: Essays toward a Church of the Future as the Organization of Philanthropy* (London: Chapman).
93. J. McCabe, *Life and Letters of George Jacob Holyoake*, Vol. 1, pp. 162–63.
94. H. Spencer to G. Holyoake (17 September 1894), NCA.
95. See M. Pellegrino Sutcliffe (2014) *Victorian Radicals and Italian Democrats* (London; New York: The Royal Historical Society, published by the Boydell Press).
96. E. Royle, *Victorian Infidels*, pp. 154–55.
97. 'On the Word "Atheist"' (25 June 1851) *Reasoner* 11.6, 88.
98. See M. Finn (1993) *After Chartism: Class and Nation in English Radical Politics, 1848–1874* (Cambridge: Cambridge University Press), esp. pp. 142–87.
99. See E. Royle, *Victorian Infidels*, p. 154.
100. R. Ashton, *142 Strand*; for Holyoake, see esp. pp. 8–9. Another overlapping circle centered around W. J. Fox and the Unitarian South Place Chapel. See B. Taylor (1993) *Eve and the New Jerusalem: Socialism and Feminism in the Nineteenth Century* (Cambridge, MA: Harvard University Press), pp. 60–74.
101. P. White (2003) *Thomas Huxley: Making the 'Man of Science'* (Cambridge: Cambridge University Press), p. 70.
102. [G. Eliot] (October 1850) 'Mackay's Progress of the Intellect', *Westminster Review* 54.2, 353–68.
103. G. J. Holyoake to H. Martineau (19 October 1851) Harriet Martineau Papers, HM 1080, University of Birmingham.
104. M. Pickering (1993) *Auguste Comte: An Intellectual Biography* (Cambridge: Cambridge University Press), pp. 435–38.
105. E. Royle, *Victorian Infidels*, p. 156.
106. Later, Holyoake claimed that Comte suggested that he had adopted the phrase from Louis Napoleon. See Holyoake and Bradlaugh, *Secularism, Scepticism, and Atheism*, pp. iv and 54.
107. Harriet Martineau, *Boston Liberator* (November 1853), quoted in the *Reasoner* 16.1 (1 January 1854), p. 5. The quote circulated widely and was found as far afield as the *Scripture Reader's Journal* for April 1856, pp. 363–64.

108. E. Syme (1 July 1853) 'Contemporary Literature of England', *The Westminster* pp. 246–88. The article was a review of several books, including of the debate between Reverend Brewin Grant and Holyoake as recorded in B. Grant and G. Holyoake (1853) *Christianity and Secularism. Report of a Public Discussion between Brewin Grant and George Jacob Holyoake, Esq. Held in the Royal British Institution, London, Commencing Jan. 20 and Ending Feb. 24, 1853* (London: Ward).
109. W. Binn (1862) 'The Religious Heresies of the Working Classes', *Westminster Review* 77, 32–52.
110. H. Mann (1854) *Census of Great Britain, 1851: Religious Worship in England and Wales* (London: G. Routledge), p. 93, emphasis in original.
111. A. Desmond (1997) *Huxley: From Devil's Disciple to Evolution's High Priest* (Reading, MA: Addison-Wesley), p. 160.
112. B. Lightman (2002) 'Huxley and Scientific Agnosticism: The Strange History of a Failed Rhetorical Strategy', *British Journal for the History of Science* 35.3, 271–89, at 284.
113. G. Holyoake (1905) *Bygones Worth Remembering* (London: T.F. Unwin), p. 64.
114. G. Eliot quoted in E. Simcox (May 1881) 'George Eliot', *The Nineteenth Century* 9, 787; E. Simcox quoted in J. Hume Clapperton (1885) *Scientific Meliorism and the Evolution of Happiness* (London: K. Paul, Trench & Co.), pp. vii–viii.
115. W. Mills, A. P. Stone, A. Wilson, and J. Redfoord Bulwer (1878) *The Law Reports*, Vol. 3 (London: Printed for the Inc. Council of Law Reporting for England and Wales, by W. Clowes and Sons), p. 607.
116. C. Knowlton, C. Bradlaugh, and A. Besant (1877) *Fruits of Philosophy: An Essay on the Population Question* (Rotterdam: v.d. Hoeven & Buys), p. vi.
117. C. Knowlton, C. Bradlaugh, and A. Besant (1877) *Fruits of Philosophy*, pp. 9–11.
118. C. Knowlton, C. Bradlaugh, and A. Besant (1877) *Fruits of Philosophy*, p. iii.
119. See, for example, the advertisement 'Books on Free Inquiry' (1854) *Reasoner* Vol. 17, pp. 95 and 256.
120. A. Besant (1885) 'Autobiographical Sketches', *Our Corner: A Monthly Magazine of Fiction, Poetry, Politics, Science, Art, Literature*, Vol. 5, p. 83.
121. Besant, 'Autobiographical Sketches', p. 31.
122. J. Green and N. J. Karolides (2005) *Encyclopedia of Censorship* (New York: Facts on File), p. 232.
123. See G. Dawson (2007) *Darwin, Literature and Victorian Respectability* (Cambridge: Cambridge University Press), pp. 116–61.
124. Besant, 'Autobiographical Sketches', p. 81.
125. Besant, 'Autobiographical Sketches', p. 80.
126. Besant, 'Autobiographical Sketches', p. 82.
127. Besant, 'Autobiographical Sketches', p. 133.
128. E. Royle, *Radicals, Secularists, and Republicans*, p. 18. The final division of the Secularists camps as a result of the Knowlton affair is at quite odds with L. Schwartz's assertion in *Infidel Feminism*, p. 200, that Holyoake 'remained neutral on the question' of the republication and defense of the *Fruits of Philosophy*. In fact, Holyoake wrote specifically to disavow the text in the press and seceded from the NSS to form a new secular union, the British Secular Union (BSU) in the aftermath of the controversy.

129. G. Dawson, *Darwin, Literature and Victorian Respectability*, pp. 116–61; and M. Mason (1994) *The Making of Victorian Sexual Attitudes* (Oxford: Oxford University Press).
130. G. Dawson, *Darwin, Literature and Victorian Respectability*, p. 119.
131. E. Royle, *Radicals, Secularists, and Republicans*, p. 120.
132. See, for example, B. Grant and G. Holyoake, *Christianity and Secularism*, pp. 56 and 200.
133. B. Grant and G. Holyoake, *Christianity and Secularism*, p. 8.
134. 'Our Policy' (1854) *The London Investigator* [afterwards the *Investigator*]: *A Monthly Journal of Secularism* 1, 1, emphasis in original.
135. Further, Robert Cooper agitated in the *Investigator* for a national Secular organization, which Holyoake, having an apparent distaste for centralization, was never quite willing to attempt. Instead, he had managed to establish a loose federation of provincial Secular societies, with London as the central headquarters, a strong Central Board, his own definition of Secularism as the doctrine, and the *Reasoner* as its official paper. See E. Royle, *Victorian Infidels*, pp. 174–77.
136. C. Southwell (1854) 'Encouraging Communications', *The London Investigator* 1, 27.
137. J. E. H. Courtney (1920) *Freethinkers of the Nineteenth Century* (New York: E.P. Dutton), p. 105.
138. H. B. Bonner and J. M. Robertson (1895) *Charles Bradlaugh: A Record of His Life and Work* (London: TF Unwin), pp. 128–30.
139. G. J. Holyoake and C. Bradlaugh, *Secularism, Scepticism, and Atheism*, p. 10.
140. G. J. Holyoake and C. Bradlaugh, *Secularism, Scepticism, and Atheism*, p. vii.
141. G. J. Holyoake and C. Bradlaugh, *Secularism, Scepticism, and Atheism*, pp. 19–20.
142. G. J. Holyoake and C. Bradlaugh, *Secularism, Scepticism, and Atheism*, p. 11.
143. G. J. Holyoake and C. Bradlaugh, *Secularism, Scepticism, and Atheism*, p. 19, emphasis added.
144. G. J. Holyoake and C. Bradlaugh, *Secularism, Scepticism, and Atheism*, pp. 8–9, emphasis added.
145. G. J. Holyoake and C. Bradlaugh, *Secularism, Scepticism, and Atheism*, p. 47.
146. G. J. Holyoake and C. Bradlaugh, *Secularism, Scepticism, and Atheism*, p. 56.
147. Carlile's book was arguably 'the first book in English to specify the methods of contraception, and ... the first progressive sex manual'. The essentials of the argument were published in his Volume 11 of the *Republican* as the essay 'What is Love?'. M. L. Bush and R. Carlile (1998) *What Is Love?: Richard Carlile's Philosophy of Sex*, (London: Verso), p. vii. In a lengthy introduction, Bush discusses the making, publication history, and reception of *Every Woman's Book*.
148. R. D. Owen (1875) *Moral Physiology, or, a Brief and Plain Treatise on the Population Question* (Boston: J. P. Mendum), pp. iii–iv; the first edition was published in 1846.
149. C. Knowlton, C. Bradlaugh, and A. Besant (1877) *Fruits of Philosophy: An Essay on the Population Question* (Rotterdam: v.d. Hoeven & Buys), p. iv.
150. E. Royle, *Radicals, Secularists, and Republicans*, p. 92.
151. The debates in the *Reasoner* in 1855 over G. Drysdale's *The Elements of Social Science* (1854) reveal Holyoake's equivocation.
152. M. Mason, *The Making of Victorian Sexual Attitudes*, pp. 284–85.

153. J. S. Mill, F. E. Mineka, F. E. L. Priestley, and J. M. Robson (1963) *The Collected Works of John Stuart Mill. (f.e.l. Priestley [Subsequently] J.M. Robson, General Editor.) (Vol. 12, 13. the Earlier Letters of John Stuart Mill, 1812–1848. Edited by F.E. Mineka.)*, 33 vols (Toronto: University of Toronto Press; London: Routledge & Kegan Paul), p. 741.
154. As I point out in the epilogue, however, Secularism did include the contradictory ambition of replacing religious belief and morality with secular values. This tension is explored there.
155. D. Nash suggests that such a *belief* is in fact common among contemporary sociologists and others who maintain the standard secularization thesis, regardless of empirical evidence and theoretical disputation to the contrary. See D. Nash (2004) 'Reconnecting Religion with Social and Cultural History: Secularization's Failure as a Master Narrative', *Cultural and Social History* 1, 302–25.
156. J. Habermas (2008) 'Notes on Post-Secular Society', *NPQ: New Perspectives Quarterly* 25.4, 17–29.
157. E. Royle, *Victorian Infidels*, pp. 160–62.
158. L. Schwartz, *Infidel Feminism*, p. 20.

4 Secularizing Science: Secularism and the Emergence of Scientific Naturalism

1. The best example is O. Chadwick (1975) *The Secularization of the European Mind in the Nineteenth Century* (Cambridge: Cambridge University Press). For the persistence of this view, see A. Keysar and B. A. Kosmin (2008) *Secularism & Science in the 21st Century* (Hartford, CT: Institute for the Study of Secularism in Society and Culture).
2. M. Stanley (2015) *Huxley's Church and Maxwell's Demon: From Theistic Science to Naturalistic Science* (Chicago: The University of Chicago Press), p. 2.
3. D. Martin (2008) 'Does the Advance of Science Mean Secularisation?', *Scottish Journal of Theology* 66.1, 51–63, at 61.
4. D. Martin, 'Does the Advance of Science Mean Secularisation?', p. 56.
5. B. V. Lightman (2014) 'Science at the Metaphysical Society: Defining Knowledge in the 1870s', in B. V. Lightman and M. S. Reidy (eds) *The Age of Scientific Naturalism: Tyndall and His Contemporaries* (London: Pickering & Chatto), pp. 187–206, at 193.
6. Recent studies include M. Stanley, *Huxley's Church and Maxwell's Demon*; B. V. Lightman and M. S. Reidy (eds) *The Age of Scientific Naturalism*; and G. Dawson (eds) (2014) *Victorian Scientific Naturalism: Community, Identity, Continuity* (Chicago; London: The University of Chicago Press). The last of these includes a review of the historiography on the category (pp. 11–17), as well as a bibliography of major secondary sources that treat it.
7. F. M. Turner (1993) *Contesting Cultural Authority: Essays in Victorian Intellectual Life* (Cambridge; New York, NY: Cambridge University Press), pp. 132–33. See also B. V. Lightman (1987) *The Origins of Agnosticism: Victorian Unbelief and the Limits of Knowledge* (Baltimore: Johns Hopkins University Press), p. 4.
8. Similarly, while Turner credits Thomas Carlyle for the scientific naturalists' emphasis on moral discipline and temperament, it is just as conceivable

that the self-disciplined, self-improvement tradition of artisan freethought served as a moral example.

9. J. Van Wyhe (2004) *Phrenology and the Origins of Victorian Scientific Naturalism* (Aldershot: Ashgate), p. 12.
10. A. Desmond (1987) 'Artisan Resistance and Evolution in Britain, 1819–1848', Osiris 3, 2nd Series (Chicago: The University of Chicago Press), pp. 77–110; A. Desmond (1989) *The Politics of Evolution: Morphology, Medicine, and Reform in Radical London* (Chicago: University of Chicago Press); and J. A. Secord (2000) *Victorian Sensation: The Extraordinary Publication, Reception, and Secret Authorship of Vestiges of the Natural History of Creation* (Chicago: University of Chicago Press), pp. 299–335.
11. J. Van Wyhe, *Phrenology and the Origins of Victorian Scientific Naturalism*, pp. 11–12.
12. B. Lightman (1989) 'Ideology, Evolution and Late-Victorian Agnostic Popularizers', in J. R. Moore (ed.) *History, Humanity and Evolution: Essays for John C. Greene* (Cambridge, UK; New York: Cambridge University Press), pp. 285–309; B. V. Lightman (2007) *Victorian Popularizers of Science: Designing Nature for New Audiences* (Chicago: University of Chicago Press), pp. 264–65; S. Paylor (2005) 'Edward B. Aveling: The People's Darwin', *Endeavour*, 29.2, 66–71; and E. Royle (1974) *Victorian Infidels: The Origins of the British Secularist Movement, 1791–1866* (Manchester: University of Manchester Press; Totowa, NJ: Rowman & Littlefield), pp. 149–77.
13. M. Rectenwald (2013) 'Secularism and the Cultures of Nineteenth-Century Scientific Naturalism', *British Journal for the History of Science* 46.2, 231–54.
14. J. R. Moore (1988) 'Freethought, Secularism, Agnosticism: The Case of Charles Darwin', in G. Parsons (ed.) *Religion in Victorian Britain: I, Traditions* (Manchester and New York: Manchester University Press), pp. 274–319.
15. For a discussion of the limits of knowledge set by scientific naturalism, see M. Stanley, *Huxley's Church and Maxwell's Demon*, pp. 80–99.
16. G. Dawson (2007) *Darwin, Literature and Victorian Respectability* (Cambridge: Cambridge University Press), p. 151; B. Lightman, 'Ideology, Evolution and Late-Victorian Agnostic Popularizers', p. 301; and M. Mason (1994) *The Making of Victorian Sexual Attitudes* (Oxford: Oxford University Press).
17. B. Lightman, 'Ideology, Evolution and Late-Victorian Agnostic Popularizers', pp. 287–88.
18. Holyoake expounded the principles of Secularism on numerous occasions in different publications, and the order and exact principles varied throughout. In fact, the principles themselves often overlapped within particular passages.
19. A. Desmond, 'Artisan Resistance and Evolution in Britain, 1819–1848', pp. 77–110; J. R. Moore (1998) 'Freethought, Secularism, Agnosticism: The Case of Charles Darwin', in G. Parsons (ed.) *Religion in Victorian Britain: I, Traditions*, pp. 274–319, at 284–85.
20. See Chapter 3 for a discussion of the founding of *The Oracle of Reason*, and its development under Holyoake. (Subsequently, *Oracle*).
21. [C. Southwell] (6 November 1841) *Oracle* 1.2, pp. 5–6.
22. J. A. Secord, *Victorian Sensation*, p. 311.
23. W. Chilton (20 November 1841) *Oracle* 1.3, 21–23, at 21.

24. J. Wiener (1983) *Radicalism and Freethought in Nineteenth-Century Britain: The Life of Richard Carlile* (Westport, CN: Greenwood Press), pp. 110–12.
25. G. Holyoake (1892) *Sixty Years of an Agitator's Life,* 2 vols (London: T. F. Unwin), Vol. 1, p. 142.
26. W. Chilton (19 February 1842) 'Theory of Regular Gradation', *Oracle* 1.9, 77–78.
27. W. Chilton (19 February 1842) 'Theory of Regular Gradation', *Oracle* 1.9, 78.
28. J. Secord, *Victorian Sensation,* p. 311.
29. W. Chilton (2 April 1842) 'Theory of Regular Gradation', *Oracle* 1.15, 123–25.
30. W. Chilton (9 April 1842) 'Theory of Regular Gradation', *Oracle* 1.16, 134.
31. The first installment of 'Theory of Regular Gradation' was on 6 November 1841.
32. W. Chilton (4 June 1842) 'The Cowardice and Dishonesty of Scientific Men', *Oracle* 1.24, 193–95.
33. W. Chilton (24 June 1843) 'Theory of Regular Gradation', *Oracle* 2.80, 219–21, at 220.
34. J. A. Secord (1989) 'Behind the Veil: Robert Chambers and *Vestiges*', in J. R. Moore (ed.) *History, Humanity and Evolution: Essays for John C. Greene* (Cambridge, UK; New York: Cambridge University Press), pp. 165–94, at 182–87.
35. W. Chilton (8 January 1845) 'Vestiges', *The Movement* 2, 9–12, at 12.
36. W. Chilton (1846) '"Materialism" and the Author of the "Vestiges"', *Reasoner* 1, pp.7–8. See also W. Chilton (1846) 'Anthropomorphism', *Reasoner* 1, pp. 36–37; and F. B. Barton, B.A. (1846) 'The Laws of Nature', *Reasoner* 2, 25–30.
37. W. Chilton to G. J. Holyoake (1 February 1846) National Co-operative Archive, Manchester (subsequently NCA). Here, Chilton reveals to Holyoake that he knows the name of the author of *Vestiges*.
38. B. Lightman (2002) 'Huxley and Scientific Agnosticism: The Strange History of a Failed Rhetorical Strategy', *British Journal for the History of Science* 35.3, 271–89.
39. L. Grugel (1976) *George Jacob Holyoake: A Study in the Evolution of a Victorian Radical* (Philadelphia: Porcupine Press), p. 83; and E. Royle (1971) *Radical Politics 1790–1900: Religion and Unbelief* (London: Longman), pp. 54–55.
40. See for example, (1877) 'Questionable Imputations Made by Mr. Bradlaugh', *Secular Review and Secularist* 1, 65–66; C. Watts (Sr.) (1877) 'The Late Trial', *Secular Review and Secularist* 1, 77; F. Neale (1877) 'The Knowlton Case', *Secular Review and Secularist* 1, 78; (1877) 'Extraordinary Statements Corrected', *Secular Review and Secularist* 1, 85–86; (In this item, a letter is included by G. J. Holyoake bitterly chiding Bradlaugh for claiming that he (Bradlaugh) had overseen the printing of Knowlton's *Fruits of Philosophy* at Holyoake's Fleet Street premises); F. Neale (1877) 'The N.S.S. and the Knowlton Pamphlet', *Secular Review and Secularist* 1, 93 and 142; (1877) 'Re-opening of Cleveland Hall', *Secular Review and Secularist* 1, 122–23; and F. Neale (1877) 'The British Secular Union', *Secular Review and Secularist* 1, 189.
41. T. L. Crosby (1997) *The Two Mr. Gladstones: A Study in Psychology and History* (New Haven: Yale University Press), pp. 177–78.
42. G. J. Holyoake (1859) *Principles of Secularism Briefly Explained* (London: Holyoake & Co.); G. J. Holyoake (1870) *The Principles of Secularism Illustrated* (London: Austin & Co.); and G. J. Holyoake and C. Bradlaugh (1870) *Secularism, Scepticism, and Atheism: Verbatim Report of the Proceedings of a Two*

Nights' Public Debate between Messrs. G J. Holyoake & C. Bradlaugh: Held at the New Hall of Science ... London, on the Evenings of March 10 and 11, 1870 (London: Austin).

43. T. H. Huxley (February 1889) 'Agnosticism', *The Nineteenth Century: A Monthly Review* 25.144, 169–94; T. H. Huxley (March 1889) 'Agnosticism: A Rejoinder', *The Nineteenth Century: A Monthly Review* 25.145, 481–504; and T. H. Huxley (June 1889) 'Agnosticism and Christianity', *The Nineteenth Century: A Monthly Review* 25.148, 937–64.
44. G. J. Holyoake (1896) *English Secularism: A Confession of Belief* (Chicago: Open Court Pub. Co.), pp. 45–49.
45. G. J. Holyoake, *English Secularism*, pp. 48–49.
46. G. Dawson and B. V. Lightman (2014) 'Introduction', in G. Dawson and B. V. Lightman, *Victorian Scientific Naturalism*, p. 3. See T. Dixon (2008) *The Invention of Altruism: Making Moral Meanings in Victorian Britain* (Oxford: Oxford University Press).
47. G. Holyoake (1843) 'Prospectus of the *Movement*', *Movement*; (1846) 'Preface' to the *Reasoner* 1, i.
48. *Reasoner* (1852) 12, 1.
49. G. J. Holyoake (1854) *Secularism, the Practical Philosophy of the People* (London: Holyoake & Co.), pp. 5–6.
50. M. Rectenwald (2013) 'Secularism and the Cultures of Nineteenth-Century Scientific Naturalism', p. 234; and J. Marsh (1998) *Word Crimes: Blasphemy, Culture, and Literature in Nineteenth-Century England* (Chicago: University of Chicago Press), p. 240.
51. *Reasoner* (1858) 23, 81.
52. G. J. Holyoake, *English Secularism*, pp. 36–37.
53. *Reasoner* (1852) 12, 1.
54. G. J. Holyoake (1896) *The Origin and Nature of Secularism; Showing that Where Freethought Commonly Ends Secularism Begins* (London: Watts), p. 51.
55. *Reasoner* (1852) 12, 127, footnote.
56. *Reasoner* (1852) 12, 34.
57. K. R. Popper (1959) *The Logic of Scientific Discovery* (New York: Basic Books, Inc.).
58. *Reasoner* (1852) 12, 34.
59. G. J. Holyoake, *English Secularism*, p. 47.
60. *Reasoner* (1851) 11, 88.
61. *Reasoner* (1852) 12, 1.
62. G. J. Holyoake, *Secularism, the Practical Philosophy of the People*, p. 6, emphasis in original.
63. In an introduction to her compilation of Comte's major works, G. Lenzer (1975) *Auguste Comte and Positivism: The Essential Writings* (New York: Harper & Row), p. xxxiii, described Comte's form of materialism as an 'anticipatory conservatism'.
64. *Reasoner* (1852) 12, 1, 130.
65. *Reasoner* (1852) 12, 1.
66. *Reasone*r (1852) 12, 130.
67. G. J. Holyoake, *Secularism the Practical Philosophy of the People*, p. 5.
68. The term is from M. Stanley, *Huxley's Church and Maxwell's Demon*, p. 8.

69. F. M. Turner (1974) *Between Science and Religion: The Reaction to Scientific Naturalism in Late Victorian England* (New Haven: Yale University Press); and F. M. Turner (1978) 'The Victorian Conflict between Science and Religion: A Professional Dimension', *Isis* 69, 356–76.
70. See A. DeWitt (2013) *Moral Authority, Men of Science, and the Victorian Novel* (Cambridge and New York: Cambridge University Press), pp. 21–52.
71. B. V. Lightman, 'Huxley and Scientific Agnosticism', esp. 280–82; and B. Cooke (2003) *The Gathering of Infidels: A Hundred Years of the Rationalist Press Association* (Amherst, NY: Prometheus Books), p. 14. The *Agnostic Annual* was to become the *RPA Annual*, the *Rationalist Annual*, and finally *Question*. See the subsequent section of this chapter for a fuller discussion of this episode.
72. T. Lewis, quoted in G. Dawson and B. V. Lightman, 'Introduction', *Victorian Scientific Naturalism*, pp. 2–7. We should be cautious in our use of such origin stories as Dawson and Lightman posit, however, as it is quite possible that the term was used elsewhere or earlier than in the American evangelical press.
73. G. Dawson and B. V. Lightman, 'Introduction', *Victorian Scientific Naturalism*, p. 6.
74. Holyoake relinquished proprietorship to Watts (Sr.), who joined with G. W. Foote, with whom Holyoake had published the *Secularist,* and whose antagonism of Bradlaugh Holyoake disapproved. Watts and Foote combined the two papers and launched *The Secular Review and Secularist* in June 1877. See L. Grugel, *George Jacob Holyoake*, pp. 142–43; L. Brake and M. Demoor (2009) *Dictionary of Nineteenth-Century Journalism in Great Britain and Ireland* (Gent: Academia Press; London: British Library), p. 566. William Stewart Ross, following Charles Albert Watts's lead, pushed the Holyoake brand of Secularism towards the new agnosticism. In January 1885, William Stewart Ross took over the *Secular Review* and gave it a new subtitle, *A Journal of Agnosticism*. Four years later he renamed it the *Agnostic Journal and Secular Review*. See B. V. Lightman, 'Huxley and Scientific Agnosticism', p. 284, and the discussion of Charles Albert Watts, below.
75. 'Naturalist', 'The Plea of a Convert', *The Secular Review and Secularist* (1878) 2, 45, emphasis added.
76. *Secular Review and Secularist* (1878) 2, 110, emphasis added.
77. J. Tyndall to G. J. Holyoake (9 November 1876) NCA.
78. G. Dawson and B. V. Lightman, 'Introduction', *Victorian Scientific Naturalism*, pp. 7–8; Huxley quoted in Dawson and Lightman.
79. T. H. Huxley (1893) 'Prologue', in *Essays upon Some Controverted Questions* (New York: D. Appleton and Co.), pp. 26–27, emphasis mine.
80. T. H. Huxley, 'Prologue', pp. 27–28.
81. M. Stanley, *Huxley's Church and Maxwell's Demon*, p. 7.
82. T. H. Huxley, 'Prologue', p. 28.
83. G. Dawson (2007) *Darwin, Literature and Victorian Respectability*, p. 120.
84. A. Desmond (1997) *Huxley: From Devil's Disciple to Evolution's High Priest* (Reading, MA: Addison-Wesley), p. 160.
85. Here I am recognizing the divergent views of Desmond and P. White (2003) *Thomas Huxley: Making the 'Man of Science'* (Cambridge; New York: Cambridge University Press); Desmond figures Huxley as a champion of industrial, middle-class values, while White sees him as working to construct science as part of an elite culture that stood in judgment of middle-class values.

86. T. H. Huxley to G. J. Holyoake (2 April 1873), NCA; J. Tyndall to G. J. Holyoake (19 November 1876), NCA; G. J. Holyoake to T. H. Huxley (20 April 1887), T. H. Huxley Papers, London, Imperial College (subsequently HP); T. H. Huxley to G. J. Holyoake (31 March 1891), NCA.
87. G. J. Holyoake to T. H. Huxley (26 March 1891), HP; H. Spencer to G. J. Holyoake (22 April 1860), NCA.
88. H. Spencer to G. J. Holyoake (14 July 1879), NCA; J. Tyndall to G. J. Holyoake (18 June 1883), NCA; G. J. Holyoake to T. H. Huxley (26 March 1891), HP.
89. T. H. Huxley to G. J. Holyoake (2 November 1875), NCA; H. Spencer to G. J. Holyoake (28 April 1875), NCA; J. Tyndall to E. Bell (15 April 1875), NCA.
90. G. Dawson, *Darwin, Literature and Victorian Respectability*, p. 119.
91. G. H. Taylor (1957) *A Chronology of British Secularism* (London: National Secular Society), p. 4.
92. G. J. Holyoake and C. Bradlaugh (1870) *Secularism, Scepticism, and Atheism: Verbatim Report of the Proceedings of a Two Nights' Public Debate between Messrs. G. J. Holyoake & C. Bradlaugh*; and G. J. Holyoake (6 August 1876) 'The Field of Action', *Secular Review* 1, 1.
93. E. Royle (1974) *Victorian Infidels*, pp. 91–92.
94. P. White, *Thomas Huxley*, p. 90; and Adrian Desmond, *Huxley*, pp. 232 and 319–21.
95. T. H. Huxley to G. J. Holyoake (2 April 1873), NCA.
96. B. Lightman, 'Ideology, Evolution and Late-Victorian Agnostic Popularizers', p. 264.
97. The letter was likely identical to the one Holyoake sent to the *Times* and the *Daily News*, in which, contrary to Bradlaugh's assertions in court, Holyoake denied having ever published Knowlton's *Fruits of Philosophy*. See *Secular Review and Secularist* (1877) 1, 43, where the letter is reprinted.
98. J. Tyndall to G. J. Holyoake (21 July 1877), NCA.
99. J. Tyndall (1915) *Fragments of Science: A Series of Detached Essays, Addresses, and Reviews*, 2 vols, Vol. 2 (New York and London: D. Appleton and Company), p. 366.
100. H. V. Mayer (1877) 'Professor Tyndall and Mr. G.J. Holyoake', *Secular Review and Secularist* 1, 293.
101. See, for example, G. J. Holyoake (27 September 1867) 'Science – The British Association for the Advancement of Science', *New York Tribune* (1866–1899), 2; and G. J. Holyoake (1 November 1868) 'The Priesthood of Science: Their Visit to Norwich', *Reasoner Review* 8, 8.
102. G. J. Holyoake to T. H. Huxley (21 June 1871), NCA.
103. R. J. Hinton (1875) *English Radical Leaders* (New York: G. P. Putnam's Sons), pp. 71–72.
104. G. J. Holyoake (27 September 1867) 'Science – The British Association for the Advancement of Science', *New York Tribune*, 2.
105. H. Spencer to G. J. Holyoake (28 April 1875), NCA.
106. J. Tyndall to E. Bell (15 April 1875), NCA.
107. T. Huxley to G. J. Holyoake (2 November 1875), NCA.
108. See for example, (1877) 'Questionable Imputations Made by Mr. Bradlaugh', *Secular Review and Secularist* 1, 65–66.
109. B. Grant and G. J. Holyoake (1853) *Christianity and Secularism. Report of a Public Discussion between Brewin Grant and George Jacob Holyoake, Esq. Held*

in the Royal British Institution, London, Commencing Jan. 20 and Ending Feb. 24, 1853 (London: Ward).
110. G. J. Holyoake to T. H. Huxley (20 April 1887), HP.
111. T. H. Huxley to G. J. Holyoake (31 March 1891), NCA.
112. G. J. Holyoake (1894) 'Characteristics of Prof. Tyndall', in H. L. Green (ed.), *John Tyndall Memorial* (Buffalo, NY: H. L. Green), pp. 1–5, at 2.
113. A. Desmond, *Huxley*, p. 501.
114. J. McCabe (1908) *Life and Letters of George Jacob Holyoake*, 2 vols (London: Watts & Co.), Vol. 2, p. 74.
115. J. R. Moore, 'Freethought, Secularism, Agnosticism', pp. 303–04.
116. *The Index* (1873) 4, p. 89.
117. See J. Green and N. J. Karolides (2005) *Encyclopedia of Censorship* (New York: Fact on File), p. 186.
118. B. Lightman, 'Ideology, Evolution and Late-Victorian Agnostic Popularizers', p. 286.
119. B. Cooke, *The Gathering of Infidels*, p. 14. *The Agnostic Annual* was to become the *RPA Annual*, the *Rationalist Annual*, and finally, *Question*.
120. L. Brake and M. Demoor, *Dictionary of Nineteenth-Century Journalism*, pp. 8–9; and B. Cooke, *The Gathering of Infidels*, p. 12.
121. B. Lightman, 'Ideology, Evolution and Late-Victorian Agnostic Popularizers'; B. Lightman, *Victorian Popularizers*, pp. 264–65; and B. Cooke, *The Gathering of Infidels*, pp. 5–29.
122. S. Paylor, 'Edward B. Aveling'.
123. E. Royle (1980) *Radicals, Secularists, and Republicans: Popular Freethought in Britain, 1866–1915* (Manchester: Manchester University Press; Totowa, NJ: Rowman & Littlefield), p. 165.
124. B. Lightman, 'Ideology, Evolution and Late-Victorian Agnostic Popularizers'; B. Lightman, *Victorian Popularizers*, pp. 264–65; and B. Cooke, *The Gathering of Infidels*, pp. 5–29.
125. E. Royle, *Radicals, Secularists, and Republicans*, p. 165.
126. For a list of RPA publications including reprints and original publications, see B. Cooke, *The Gathering of Infidels*, Appendix 1, pp. 305–17.
127. A. Desmond, *Huxley*, p. 527; and B. V. Lightman, 'Huxley and Scientific Agnosticism', pp. 280–82.
128. B. Cooke, *The Gathering of Infidels*, pp. 12–13, 38. Leonard Huxley was listed among the Honorary Associates of the RPA in the *Agnostic Annual and Ethical Review* (1907), p. 82.
129. Quoted in B. Cooke, *The Gathering of Infidels*, p. 12.
130. In late 1883, Huxley wrote to Tyndall about Watts's 'impudence' for 'printing this without asking leave or sending a proof, but paraded me as a "contributor"'. T. H. Huxley to J. Tyndall (25 November 1883), Tyndall Papers; B. V. Lightman, *Victorian Popularizers*, p. 264.
131. B. Cooke, *The Gathering of Infidels*, pp. 12–13.
132. T. H. Huxley (1892) 'Possibilities and Impossibilities', *The Agnostic Annual*, 3–10; and J. V. Jensen (1991) *Thomas Henry Huxley: Communicating for Science* (Newark: University of Delaware Press; London: Associated University Presses), p. 122, also agrees that Huxley voluntarily published the essay in the *Agnostic Annual*.
133. E. Royle, *Radicals, Secularists and Republicans*, p. 166.

134. B. Cooke, *The Gathering of Infidels*, p. 38.
135. A. Desmond, *Huxley*, pp. 527–28.
136. B. Cooke, *The Gathering of Infidels*, pp. 12–13, 38.
137. A. Desmond, *Huxley*, p. 580; and B. Cooke, *The Gathering of Infidels*, Appendix 2, p. 318.
138. T. H. Huxley to G. J. Holyoake (9 May 1884), NCA.
139. T. H. Huxley to G. J. Holyoake (2 April 1873), NCA; G. J. Holyoake to T. H. Huxley (20 April 1887), HP; T. H. Huxley to G. J. Holyoake (26 April 1887); J. Tyndall to G. J. Holyoake (16 November 1876); J. Tyndall to G. J. Holyoake (21 July 1877), NCA; H. Spencer to G. J. Holyoake (14 July 1879), NCA.
140. G. Dawson and B. V. Lightman, 'Introduction', *Victorian Scientific Naturalism*, pp. 14–15.
141. M. Stanley, *Huxley's Church and Maxwell's Demon*, passim.

5 The Three Newmans: A Triumvirate of Secularity

1. Anon. (1851) 'Forms of Infidelity in the Nineteenth Century', *The North British Review* 15.29, 35–56, at 52.
2. W. Robbins (1966) *The Newman Brothers: An Essay in Comparative Intellectual Biography* (London: Heinemann Educational Books).
3. B. Willey (1956) 'Francis W. Newman', *More Nineteenth-Century Studies: A Group of Honest Doubters* (New York: Harper & Row), p. 11; hereafter referred to as *Honest Doubters*.
4. J. M. Wheeler (1891) 'Biographical Sketch', in C. R. Newman, G. J. Holyoake, and J. M. Wheeler (eds) *Essays on Rationalism: By Charles Robert Newman ... With Preface by George Jacob Holyoake. And Biographical Sketch by J.M. Wheeler* (London: Progressive Pub. Co.), p. 9.
5. A notable exception is J. M. Svaglic (1956) 'Charles Newman and His Brothers', *PMLA* 71.3, 370–85.
6. J. M. Wheeler, 'Biographical Sketch', p. 9.
7. J. M. Wheeler, 'Biographical Sketch', p. 15.
8. C. R. Newman, *Essays in Rationalism*.
9. D. Hempton (2008) *Evangelical Disenchantment: Nine Portraits of Faith and Doubt* (New Haven and London: Yale University Press), p. 13.
10. The only book-length treatments of Francis Newman, other than an unsympathetic and damaging biography by I. Giberne Sieveking in 1909, are two unpublished dissertations. See Ann Margaret Schellenberg (1994) 'Prize the Doubt: The Life and Work of Francis William Newman', unpublished doctoral dissertation, University of Durham; and J. R. Bennett (1961) 'Francis W. Newman: Religious Liberalism in Nineteenth-Century England', unpublished doctoral dissertation, Stanford University. In the first chapter of his dissertation, 'One Hundred Years of Criticism of Newman', Bennett discusses Newman's prominent (and often vitriolic) critics, who included Sieveking, Henry Rogers, Matthew Arnold, and Lionel Trilling, among many others. As discussed below, Newman was also thrashed in the periodical press by numerous and sundry anonymous critics. Bennett also cites Newman's most prominent supporters, including John Sterling, James Martineau, George Eliot, George Holyoake, Alfred William Benn, and Basil Willey. Periodical

supporters included the *Westminster Review*, the *Reasoner*, and the *Leader*, to all of which Newman contributed. Sieveking's biography, as both Willey and Bennett note, is worth citing only for his own words and as a specimen of the kind of ire his critique of dominant Christianity incited.

11. Many of his other, often later concerns are much more readily comprehensible. These include advocacy for women's rights, vegetarianism, abolitionism, teetotalism, anti-vivisection, and his anti-vaccination position, among others.
12. See E. Royle (1974) *Victorian Infidels: The Origins of the British Secularist Movement 1791–1866* (Manchester: University of Manchester Press; Totowa, NJ: Rowman & Littlefield), pp. 155–58.
13. The phrase is from Leigh Hunt's book: L. Hunt (1853) *The Religion of the Heart: A Manual of Faith and Duty* (London: John Chapman). But the phrase aptly describes the school that Francis Newman effectively inaugurated with F. W. Newman (1849) *The Soul: Its Sorrows and Aspirations: An Essay towards the Natural History of the Soul, as the True Basis of Theology* (London: Chapman). Hunt approvingly referred to *The Soul* in *The Religion of the Heart* (p. 84). A 'religion of the heart' could also be found much earlier in evangelical persuasions of traditional Christianity. However, the phrase refers here to a justification of faith by the heart or soul alone, independent of, or even *as*, a doctrine.
14. Pietro Corsi uses this phrase to describe the religious school of Newman and others in P. Corsi (1988) *Science and Religion: Baden Powell and the Anglican Debate, 1800–1860* (Cambridge, UK; New York: Cambridge University Press), p. 200.
15. This phrase was used by Newman himself in F. Newman (1862) *The Soul: Its Sorrows and Aspirations: An Essay towards the Natural History of the Soul, as the True Basis of Theology* (London: George Manwaring), p. 183. Hereafter *The Soul*. The original publication date was 1849. Unless otherwise noted, subsequent citations of the text will be to this, the seventh edition.
16. Newman's dismissal of miracles, superstition, the inerrancy of the Bible, and the divinity of Jesus led some to label him as a rationalist or rational religionist, although Newman disavowed the moniker of rationalism as such. See discussion to follow.
17. See discussion in the conclusion to this chapter.
18. A. W. Benn (1906) *The History of English Rationalism in the Nineteenth Century*, 2 vols, Vol. 2 (New York and Bombay: Longmans, Green, & Co.), pp. 18 and 28. As P. Corsi points out, 'At the opposite side of the theological spectrum [from the Tractarians], the "mystics" – a term employed by Baden Powell to indicate the followers of the American transcendental school, or individual thinkers like Blanco White, Francis Newman, John Daniel Morell (1816–1891) or John Sterling (1806–1844) – made religious ideas and feelings the basis of their philosophy of religion'. See P. Corsi, *Science and Religion*, p. 194.
19. See T. Larsen (2001) 'The Regaining of Faith: Reconversions among Popular Radicals in Mid-Victorian England', *Church History* 70.3, 527–43, at 534.
20. J. R. Moore (1988) 'Freethought, Secularism, Agnosticism: The Case of Charles Darwin', in G. Parsons (ed.) *Religion in Victorian Britain: I, Traditions*, (Manchester and New York: Manchester University Press), p. 298.
21. The word 'husk' is used by Basil Willey to refer to Newman's shedding of layer after layer of religious doctrine as recounted in *Phases of Faith*; see B. Willey, *Honest Doubters*, p. 12.

22. J. H. Newman (1870) *An Essay in Aid of a Grammar of Assent* (New York: The Catholic Publication Society), p. 234; hereafter, *Grammar of Assent*.
23. J. H. Newman, *Grammar of Assent*, p. 234.
24. J. H. Newman, *Grammar of Assent*, pp. 234–35.
25. J. H. Newman, *Grammar of Assent*, 235.
26. J. H. Newman, *Grammar of Assent*, 235–36.
27. Quoted in M. Ward (1948) *Young Mr. Newman* (New York: Sheed & Ward), p. 360.
28. B. Willey, *Honest Doubters*, p. 11.
29. See M. Ward, *Young Mr. Newman*, pp. 360–61.
30. D. Hempton, *Evangelical Disenchantment*, p. 4.
31. For reasons that adherents might leave evangelicalism, see D. Hempton, *Evangelical Disenchantment*, pp. 190–93; quote at 192.
32. D. Hempton, *Evangelical Disenchantment*, p. 196.
33. F. M. Turner (2002) *John Henry Newman: The Challenge to Evangelical Religion* (New Haven & London: Yale University Press), p. 23.
34. F. M. Turner, *John Henry Newman*, p. 11.
35. W. R. Ward (2006) *Early Evangelicalism: A Global Intellectual History, 1670–1789* (Cambridge: Cambridge University Press).
36. T. Purnell (1884) 'Charles Robert Newman', *The Athenaeum: Journal of Literature, Science, The Fine Arts, Music and the Drama*, No. 2944, p. 408; J. M. Wheeler (1891) 'Biographical Sketch'; S. O'Faoláin (1952) *Newman's Way* (London: Longmans, Green & Co.); M. Ward, *Young Mr. Newman*; and J. M. Svaglic, 'Charles Newman and His Brothers'.
37. J. M. Wheeler (1891) 'Biographical Sketch', pp. 10–11; and J. M. Svaglic, 'Charles Newman and His Brothers', p. 375.
38. Quoted in J. M. Svaglic, 'Charles Newman and His Brothers', p. 376.
39. J. M. Svaglic, 'Charles Newman and His Brothers', p. 376.
40. T. Larsen, 'The Regaining of Faith', pp. 541–42.
41. A. W. Benn, *The History of English Rationalism*, Vol. 2, pp. 26–7.
42. B. Willey, *Honest Doubters*, p. 18.
43. W. Robbins, *The Newman Brothers*, p. xi.
44. K. Manwaring (1988) 'The Forgotten Brother: Francis William Newman, Victorian Modernist', *Courier* 23.1, 3–26, at 26.
45. A. W. Benn, *The History of English Rationalism*, Vol. 2, p. 18, emphasis mine.
46. Letter from F. W. Newman to G. Griffen (16 January 1860) quoted in M. Francis (2007) *Herbert Spencer and the Invention of Modern Life* (Ithaca, NY: Cornell University Press), p. 368, note 54.
47. W. F. Bynum (2000) 'The Cardinal's Brother: Francis Newman, Victorian Bourgeois', in P. Gay, M. S. Micale, and R. L. Dietle (eds) *Enlightenment, Passion, Modernity: Historical Essays in European Thought and Culture* (Stanford, CA: Stanford University Press), pp. 131–47, at 132, emphasis mine.
48. W. F. Bynum, 'The Cardinal's Brother', p. 131, emphasis mine.
49. R. H. Ross and A. Tennyson (1973) *Alfred, Lord Tennyson in Memoriam: An Authoritative Text Backgrounds and Sources Criticism* (New York: Norton), section 96, p. 62.
50. F. W. Newman (1850) *Phases of Faith; or, Passages from the History of My Creed* (London: J. Chapman), p. iii.
51. D. Hempton, *Evangelical Disenchantment*, p. 42.

52. Quoted in D. Hempton, *Evangelical Disenchantment*, p. 43.
53. Manwaring, 'The Forgotten Brother', p. 6.
54. F. W. Newman, *Phases of Faith*, p. 1.
55. F. M. Turner (2002) *John Henry Newman*, p. 112.
56. F. W. Newman, *Phases of Faith*, p. 14.
57. S. L. Stephen, R. Blake, and C. S. Nicholls (1921) *The Dictionary of National Biography* (London: Oxford University Press), Vol. 22, p. 193.
58. B. Willey, *Honest Doubters*, p. 18; and W. F. Bynum, 'The Cardinal's Brother', p. 49.
59. F. M. Turner points out in *John Henry Newman*, pp. 48–49, 'Darby had already begun to develop his particular premillennial views of prophecy, involving the concept of the Rapture. His ideas had enormous influence later in the century throughout the world of prophecy study, especially in the United States, where his ideas constituted the framework for the Schofield Reference Bible, a key text to modern American fundamentalism'.
60. F. M. Turner, *John Henry Newman*, p. 48; and *Littell's Living Age* (1885) Fifth Series, Vol. 52, 350.
61. W. F. Bynum, 'The Cardinal's Brother', p. 139.
62. The missionary trip is recounted in Newman's latterly published book, F. W. Newman (1856) *Personal Narrative, in Letters, Principally From Turkey* (London: Holyoake and Co.). The best brief account of this disastrous missionary pilgrimage and its impact on Newman's life and belief is in D. Hempton, *Evangelical Disenchantment*, pp. 45–51.
63. F. W. Newman, *Phases of Faith*, pp. 52–53.
64. F. W. Newman, *Phases of Faith*, p. 57.
65. F. W. Newman, *Phases of Faith*, pp. 58–59.
66. F. W. Newman, *Phases of Faith*, pp. 59–60.
67. A. W. Benn, *The History of English Rationalism*, Vol. 2, pp. 19–20.
68. F. W. Newman, *Phases of Faith*, p. 73.
69. F. W. Newman, *Phases of Faith*, p. 72.
70. G. J. Holyoake (1851) *The Philosophic Type of Religion: As Developed by Prof. Newman, Stated, Examined, and Answered* (London: J. Watson), p. 6.
71. F. W. Newman, *Phases of Faith*, p. 76.
72. F. W. Newman, *Phases of Faith*, pp. 78–79.
73. F. W. Newman, *Phases of Faith*, pp. 112 and 122.
74. F. W. Newman, *Phases of Faith*, p. 107.
75. F. W. Newman, *Phases of Faith*, pp. 110 and 136.
76. F. W. Newman, *Phases of Faith*, p. 135.
77. F. W. Newman, *Phases of Faith*, pp. 139–41.
78. W. R. Ward, *Early Evangelicalism*, p. 3.
79. F. W. Newman, *Phases of Faith*, pp. 117–18, emphasis mine.
80. F. W. Newman, *Phases of Faith*, p. 117.
81. F. W. Newman, *Phases of Faith*, pp. 118–19.
82. F. W. Newman, *Phases of Faith*, p. 118.
83. F. W. Newman, *Phases of Faith*, p. 119.
84. F. W. Newman, *Phases of Faith*, p. 159.
85. F. W. Newman, *Phases of Faith*, p. 155, emphasis in original.
86. F. W. Newman, *Phases of Faith*, p. 187.
87. F. W. Newman, *Phases of Faith*, p. 188, emphasis in original.
88. F. W. Newman, *Phases of Faith*, p. 188.

89. F. W. Newman, *Phases of Faith*, p. 220.
90. F. W. Newman, *Phases of Faith*, p. 221.
91. F. W. Newman, *Phases of Faith*, p. 232.
92. T. Larson, 'The Regaining of Faith', p. 533.
93. F. W. Newman, *Phases of Faith*, p. 233.
94. In this Newman followed a trend in de-conversions from evangelicalism. See D. Hempton, *Evangelical Disenchantment*, p. 191.
95. W. H. Dunn (1961) *James Anthony Froude, a Biography* (Oxford: Clarendon Press), p. 134, note 1.
96. Anon. (1850) 'Newman's Phases of Faith', *The British Quarterly Review* 12.23, 1–56, at 56.
97. Anon. (1851) 'Forms of Infidelity in the Nineteenth Century', *The North British Review* 15.29, 47–48.
98. Anon. (1851) 'Modern "Spiritualism"', *The Journal of Sacred Literature* 7.14, 360–77.
99. Anon. (1851) 'Newman on the True Basis of Theology', *Brownson's Quarterly Review* n.s. 5.4, 417.
100. D. Walther (1851) *Some Reply to 'Phases of Faith'* (London: J. K. Campbell), p. 3, emphasis in original.
101. [J. N. Darby] (1853) *The Irrationalism of Infidelity, Being a Reply to 'Phases of Faith'* (London: Groombridge and Sons), pp. v–vi.
102. See J. A. Secord (2000) *Victorian Sensation: The Extraordinary Publication, Reception, and Secret Authorship of Vestiges of the Natural History of Creation* (Chicago: University of Chicago Press).
103. F. W. Newman, *The Soul*, p. 9.
104. D. Hempton, *Evangelical Disenchantment*, p. 59.
105. F. W. Newman, *The Soul*, p. 8, emphasis in original.
106. F. W. Newman, *The Soul*, p. 9.
107. F. W. Newman, *The Soul*, p. 10. For the influence of Auguste Comte's ideas on Newman, see C. D. Cashdollar (1989) *The Transformation of Theology, 1830–1890: Positivism and Protestant thought in Britain and America* (Princeton, NJ: Princeton University Press), pp. 312–14.
108. F. W. Newman, *The Soul*, p. 12.
109. F. W. Newman, *The Soul*, p. 14.
110. F. W. Newman, *The Soul*, p. 15.
111. F. W. Newman, *The Soul*, p. 23.
112. F. W. Newman, *The Soul*, p. 23.
113. F. W. Newman, *The Soul*, pp. 23–24.
114. F. W. Newman, *The Soul*, p. 25.
115. F. W. Newman, *The Soul*, p. 30.
116. F. W. Newman, *The Soul*, p. 39, emphasis in original.
117. F. W. Newman, *The Soul*, p. 39, emphasis in original.
118. F. W. Newman, *The Soul*, p. 78, emphasis in original.
119. F. W. Newman, *The Soul*, p. 82, emphasis in original.
120. F. W. Newman, *The Soul*, p. 85.
121. F. W. Newman, *The Soul*, p. 94.
122. S. During (2013) 'George Eliot and Secularism', in A. Anderson and H. E. Shaw (eds) *A Companion to George Eliot* (Chichester, UK; Malden, MA: Wiley-Blackwell), pp. 428–41, at 435.
123. D. Hempton, *Evangelical Disenchantment*, p. 62.

124. F. W. Newman, *The Soul*, p. 154, emphasis in original.
125. F. W. Newman, *The Soul*, p. 154, emphasis in original.
126. D. Hempton, *Evangelical Disenchantment*, p. 63.
127. F. W. Newman, *The Soul*, p. 156.
128. F. W. Newman, *The Soul*, p. 158.
129. See S. J. Gould (1999) *Rocks of Ages: Science and Religion in the Fullness of Life* (New York: Ballantine Pub. Group).
130. G. J. Holyoake, 'The Philosophic Type of Religion, Developed by Prof. Newman: Stated and Examined', *Reasoner* 11.6, 83–86; [William H. Ashurst] (25 June 1851) 'On the Word "Atheist"', *Reasoner* 11.6, 88, emphasis in original. In response to Ashurst, Holyoake proposed the term '*Secularism*' to refer to 'the work we have always had in hand, and how it is larger than Atheism'.
131. G. J. Holyoake, *The Philosophic Type of Religion*.
132. Quoted in '"Infidel" Progress', *Reasoner* (15 October 1854) Vol. 17.16, 252.
133. S. D. Collet (1855) *George Jacob Holyoake and Modern Atheism: A Biographical Essay* (London: Trübner & Co.), pp. 18–20.
134. M. Francis (2007) *Herbert Spencer and the Invention of Modern Life* (Ithaca: Cornell University Press), p. 115.
135. M. Francis, *Herbert Spencer*, p. 114.
136. M. Francis, *Herbert Spencer*, p. 115. The term 'spiritualist' did not have the meaning it would take by the end of the century. It referred rather to a believer of the Francis Newman kind.
137. See F. W. Newman (1854) *Catholic Union: Essays towards a Church of the Future, as the Organization of Philanthropy* (London: J. Chapman). Secularists are mentioned explicitly on page 99: 'I mean, many who call themselves Socialists and Secularists, of whom the latter, under their able, upright and estimable leader, G. J. Holyoake, have already attained a considerable organization, and would at once be valuable allies.
138. S. D. Collet, *George Jacob Holyoake and Modern Atheism*, p. 24.
139. S. D. Collet, *George Jacob Holyoake and Modern Atheism*, p. 21.
140. F. W. Newman to J. H. Newman (17 January 1860) quoted in E. Short (2013) *Newman and His Family* (London: Bloomsbury), pp. 191–92.
141. G. J. Holyoake (1851) *The Last Days of Mrs. Emma Martin, Advocate of Freethought* (London: Watson) quoted in S. Dobson Collet, *George Jacob Holyoake and Modern Atheism*, pp. 20–21, emphasis in original.
142. G. J. Holyoake (1905) 'The Three Newmans', in *Bygones Worth Remembering* 2 Vols (New York: E. P. Dutton and Co.), Vol. 1, pp. 192–201, at 192.

6 George Eliot: The Secular Sublime, Post-Secularism, and 'Secularization'

1. For brief summaries of Eliot's secular outlook, see N. Henry (2012) *The Life of George Eliot, A Critical Biography* (Malden, MA: Wiley-Blackwell), pp. 51–59; D. Hempton (2008) *Evangelical Disenchantment: Nine Portraits of Faith and Doubt* (New Haven and London: Yale University Press), pp. 19–40; N. Vance (2013) *Bible and Novel: Narrative Authority and the Death of God* (Oxford: Oxford University Press), pp. 93–114; S. During (2013) 'George Eliot and

Secularism', in A. Anderson and H. E. Shaw (eds) *A Companion to George Eliot* (Hoboken, NJ: John Wiley & Sons Inc.), pp. 428–41; and M. Rectenwald (2013) 'Secularism', in M. Harris (ed.) *George Eliot in Context* (Cambridge; New York: Cambridge University Press), pp. 271–78.

2. D. Coleman (2008) '*Daniel Deronda* and the Limits of the Sermonic Voice', *Studies in the Novel* 40.4, 407–25, at 407. Coleman draws the limit of the sermonic voice of Christianity at *Daniel Deronda*, where she suggests that it no longer functions as such.
3. N. Vance, *Bible and Novel*, p. 100.
4. G. Eliot and G. S. Haight (1954) *The George Eliot Letters,* 7 vols, Vol. 1 (New Haven: Yale University Press), p. 282. Eliot and Newman had become friends through the *Westminster Review* in the early 1850s, but she had been introduced to his work before meeting him. In particular, she read and was impressed by *The Soul* (1849).
5. G. Eliot and G. S. Haight, *Letters,* Vol. 3, p. 231.
6. I call these 'secular-religious' projects, because they aimed to create religions on largely secular grounds. Holyoake's Secularism may be an exception here, although he wavered on the question of whether Secularism was a religion.
7. As evidenced in her correspondence, Eliot certainly took notice of George Holyoake's work. In a letter to Sarah Hennell on 28 July 1858, Eliot remarked that however 'much [she] dislike[ed] the *Reasoner*', she considered Holyoake himself to be 'so superior a man ... who has written so well elsewhere'. See G. Eliot and G. S. Haight, *Letters,* Vol. 2, p. 473. On 13 April 1861, she wrote to Sarah Hennnell and mentioned Holyoake, noting that she thought 'Holyoake had renounced the term "Atheist" long, long ago: I have often [heard] Mr. Lewes speak of his having done so'. See *Letters*, Vol. 3, p. 404. On 1 August 1875, George Henry Lewes wrote to Holyoake, asking for a copy of his *A History of Co-operation in England* (1875–79), as 'Mrs. Lewes would *very* much like to have your book'. See *Letters*, Vol. 6, p. 160, emphasis in original. George Eliot wrote to George Holyoake on 11 August 1879, stating that 'I have been able already to read a good deal in your volume' (*A History of Co-operation*, Vol. 2), and noting that '[t]he description of the ruined school at New Lanark has an almost tragic impressiveness'. See *Letters*, Vol. 7, p. 193. It is very clear that at least Lewes read the *Reasoner* regularly, as his letter to Herbert Spencer on 9 March 1856 attests. See *Letters*, Vol. 8, p. 151.
8. F. W. Newman (1854) *Catholic Union: Essays toward a Church of the Future as the Organization of Philanthropy* (London: Chapman); and J. H. Allen (1891) *Positive Religion: Essays, Fragments, and Hints* (Boston: Roberts Bros).
9. For Comte's attempts at cooperation with religious groups, especially Catholics, see M. Pickering (1993) *Auguste Comte: An Intellectual Biography* (Cambridge: Cambridge University Press), pp. 417–23.
10. S. During (2013) 'George Eliot and Secularism', p. 428.
11. For the issue of conversion in *Daniel Deronda*, see E. W. Heady (2013) *Victorian Conversion Narratives and Reading Communities* (Burlington, VT: Ashgate), pp. 75–103.
12. G. Eliot and G. Maertz (2004) *Middlemarch: A Study of Provincial Life* (Peterborough, ON: Broadview Press), p. 640.
13. The terms 'philological secularism' and 'philosophical secularism' are drawn from S. During, 'George Eliot and Secularism', p. 430. By philological

secularism, During means the work done by such enterprises as the Higher Criticism. By philosophical secularism, he refers to the work of Darwinism and scientific naturalism more generally.

14. G. Eliot and G. Maertz, *Middlemarch*, p. 32.
15. Dorothea is associated with the image of a nun in several places in *Middlemarch*. The novel compares her with Saint Theresa of Avila in the Prelude, and she is associated with nuns by Naumann, the German painter and friend of Will Ladislaw, and by Mrs. Cadwallader.
16. C. Taylor (2007) *A Secular Age* (Cambridge, MA: Belknap Press of Harvard University Press), p. 329.
17. For the notion of Jewishness as a 'paradigm of textual difference' (in addition to cultural, ethnic, and religious difference) in *Daniel Deronda*, see C. Scheinberg (2010) '"The Beloved Ideas Made Flesh": *Daniel Deronda* and Jewish Poetics', *ELH* 77.3, 813–39.
18. For discussions of cosmopolitanism in *Daniel Deronda*, see T. Albrecht (2012) '"The Balance of Separateness and Communication": Cosmopolitan Ethics in George Eliot's *Daniel Deronda*', *ELH* 79.2, 389–416; K. Bailey Linehan (1992) 'Mixed Politics: The Critique of Imperialism in *Daniel Deronda*', *Texas Studies in Literature and Language* 34.3, 323–46; and P. Brantlinger (1992) 'Nations and Novels: Disraeli, George Eliot, and Orientalism', *Victorian Studies* 35.3, 255–75.
19. G. Eliot (1876) *Daniel Deronda* (Edinburgh and London: William Blackwood and Sons) Vol. 2, Book 4, p. 326. Unless otherwise noted, all citations of *Daniel Deronda* will be to this edition.
20. See especially F. Bonaparte (1993) '*Daniel Deronda*: Theology in a Secular Age', *Religion & Literature* 25.3, 17–44; see also P. Brantlinger, 'Nations and Novels', 269.
21. C. Sheinberg (2010) '"The Beloved Ideas Made Flesh": *Daniel Deronda* and Jewish Poetics', *ELH* 77.3, 813–839, at 817.
22. G. Smith (2008) *A Short History of Secularism* (London: I. B. Tauris), p. 2.
23. G. Eliot, *Daniel Deronda*, Vol. 2, Book 4, p. 290.
24. W. E. Connolly (1999) *Why I am Not a Secularist* (Minneapolis [u.a.]: University of Minnesota Press), p. 9.
25. G. Eliot, *Daniel Deronda*, Vol. 2, Book 4, p. 295.
26. G. Eliot, *Daniel Deronda*, Vol. 1, Book 1, p. 67.
27. G. Eliot (1884) *Daniel Deronda* (Edinburgh and London: William Blackwood and Sons), Book 8, p. 606. All citations of Book 8 will refer to this edition.
28. George Eliot, *Daniel Deronda*, Book 8, p. 565.
29. As Jane Irwin points out, drawing from the diary of George Henry Lewes, the Gwendolen gambling scene is based on Eliot's and Lewes's trip to Homburg, where on 26 September 1874, they saw amongst the roulette players 'Miss Leigh (Byron's granddaughter) having lost 500 £, looking feverishly excited. Painful sight'. See G. Eliot and J. Irwin (1996) *George Eliot's Daniel Deronda Notebooks* (Cambridge: Cambridge University Press), p. xxvii.
30. G. Eliot, *Daniel Deronda*, Vol 1, Book 1, p. 9.
31. G. Eliot, *Daniel Deronda*, Vol 2, Book 3, p. 142.
32. For a discussion of Grandcourt as having no soul, see F. Bonaparte, '*Daniel Deronda*: Theology in a Secular Age', 26–27.
33. G. Eliot, *Daniel Deronda*, Vol. 1, Book 1, p. 31.

34. G. Eliot and G. Maertz, *Middlemarch*, p. 606.
35. W. Wordsworth (1798) 'Lines Composed a Few Miles above Tintern Abbey, on Revisiting the Banks of the Wye during a Tour. July 13, 1798', in *The Poetical Works of William Wordsworth*, 3 Vols, Vol. 2 (London: Macmillan and Co, Ltd.), pp. 52–53, emphasis mine.
36. G. Eliot, *Daniel Deronda*, Vol. 1, Book 1, p. 32.
37. G. Eliot, *Daniel Deronda*, Vol. 1, Book 1, p. 120.
38. F. Bonaparte, '*Daniel Deronda*: Theology in a Secular Age', 25–26.
39. M. Francis (2007) *Herbert Spencer and the Invention of Modern Life* (Ithaca, NY: Cornell University Press), pp. 111–31.
40. G. Eliot, *Daniel Deronda*, Vol. 1, Book 1, p. 145.
41. G. Eliot, *Daniel Deronda*, Vol. 2, Book 4, pp. 293–96.
42. G. Eliot, *Daniel Deronda*, Vol. 1, Book 1, p. 32. As Bonaparte argues, Gwendolen represents England in the novel. See F. Bonaparte, '*Daniel Deronda*: Theology in a Secular Age', 24.
43. G. Eliot, *Daniel Deronda*, Vol. 2, Book 4, p. 296.
44. G. Eliot (1879) 'The Modern Hep! Hep! Hep!', in *Impressions of Theophrastus Such* (New York: Harper & Brothers), p. 220.
45. G. Eliot, 'The Modern Hep! Hep! Hep!', p. 226.
46. G. Eliot, *Daniel Deronda*, Vol. 1, Book 1, p. 86. 'The Spoiled Child' is the title of Book 1.
47. G. Eliot, *Daniel Deronda*, Vol. 2, Book 3, p. 42.
48. G. Eliot, *Daniel Deronda*, Vo1. 1, Book 1, p. 109.
49. G. Eliot, *Daniel Deronda*, Vol. 2, Book 4, p. 290.
50. G. Eliot, *Daniel Deronda*, Vol. 2, Book 4, p. 289, emphasis in original.
51. For these issues, see M. Ragussis (1995) *Figures of Conversion: The Jewish Question & English National Identity* (Durham, NC: Duke University Press); G. Alderman (1992) *Modern British Jewry*, 2nd ed. (Oxford: Oxford University Press); T. M. Endelman (2002) *The Jews of Britain 1656–2000* (Berkeley: University of California Press); and D. S. Katz (1996) *The Jews in the History of England 1485–1850* (Oxford: Oxford University Press). Following Ragussis, A. T. Levenson notes that 'Jewish characters in English novels were often "figures of conversion". They served as the literary counterparts to the memoirs produced by the saved souls of the Society for the Conversion of the Jews'. In *Daniel Deronda*, Eliot inverts this conversion narrative. See A. T. Levenson (2008) 'Writing the Philosemitic Novel: *Daniel Deronda* Revisited', *Prooftexts* 28.2, 129–56, at 144.
52. G. Eliot, *Daniel Deronda*, Vol. 2, Book 4, p. 290.
53. G. Eliot, *Daniel Deronda*, Vol 2, Book 4, p. 316. Susan Meyer has argued that the novel 'attempts to idealize its diminished women', including the mother of Mirah and Mordecai, and Mirah herself, whose 'impossibly tiny feet' and other diminutive bodily and character features makes her 'difficult to take' for many readers. Such exaltation of subordinated women, I would argue, has to do with the novel's privileging of cultural and religious transmission over the self-emancipation and self-realization of individual characters. Of course, the exception is Daniel himself, who is able to meld his own notions of self-fulfillment with submission to the social. Likewise, Meyer's critique holds. See S. Meyer (1996) *Imperialism at Home: Race and Victorian Women's Fiction* (Ithaca, NY: Cornell University Press), pp. 173–75.

54. G. Eliot, *Daniel Deronda*, Vol 3, Book 6, p. 205.
55. Quoted in C. J. Marks (2001) '*Daniel Deronda*: Community, Spirituality, and Minor Literature', *Partisan Review*, 446–63, at 447.
56. G. Eliot, *Daniel Deronda*, Vol. 3, Book 6, p. 226.
57. G. Eliot, *Daniel Deronda*, Vol. 3, Book 6, pp. 222–55.
58. G. Eliot, *Daniel Deronda*, Vol. 3, Book 6, p. 235.
59. W. D. Rubinstein, M. Jolles, and H. L. Rubinstein (2011) *The Palgrave Dictionary of Anglo-Jewish History* (Basingstoke: Palgrave Macmillan), p. 414.
60. A. G. Henriques (1864) *On Some Legal and Economic Questions Connected with Land-Credit & Mortgage Companies* (London: Wilson).
61. As numerous scholars have noted, the Judaic Reform movement in Britain was extremely moderate in comparison with movements in Germany and the United States, for example. The West London Synagogue maintained nearly the same services as the Orthodox synagogues.
62. G. Cantor (2006) 'Anglo-Jewish Responses to Evolution', in G. Cantor and M. Swetlitz (eds) *Jewish Tradition and the Challenge of Darwinism* (Chicago: University of Chicago Press), pp. 23–46, at 37–40.
63. G. Cantor, 'Anglo-Jewish Responses to Evolution', p. 39.
64. G. Cantor, 'Anglo-Jewish Responses to Evolution,' p. 24.
65. See G. Cantor, 'Anglo-Jewish Responses to Evolution', pp. 23–24 and 40–46; See also W. D. Rubinstein, M. Jolles, and H. L. Rubinstein, *The Palgrave Dictionary of Anglo-Jewish History*, pp. 558–59.
66. G. Eliot, *Daniel Deronda*, Vol. 3, Book 6, p. 231.
67. F. Bonaparte (1993) '*Daniel Deronda*: Theology in a Secular Age', pp. 19–20.
68. G. Eliot, *Daniel Deronda*, Vol. 3, Book 5, p. 178.
69. A. Singh (2006) *Literary Secularism: Religion and Modernity in Twentieth-Century Fiction* (Newcastle: Cambridge Scholars Press), pp. 10–11, emphasis in original.
70. S. Meyer, *Imperialism at Home*, p. 177.
71. G. Eliot, *Daniel Deronda*, Vol. 3, Book 6, p. 353. The phrase refers to Gwendolen's condition under Grandcourt's oppression.
72. N. Henry, *The Life of George Eliot*, pp. 90–91. See also D. Atkins (1978) *George Eliot and Spinoza* (Salzburg: Inst. f. Engl. Sprache u. Literatur, Univ. Salzburg).
73. V. M. Nemoianu (2010) 'The Spinozist Freedom of George Eliot's *Daniel Deronda*', *Philosophy and Literature* 34.1, 65–81, at 67.
74. V. M. Nemoianu (2010) 'The Spinozist Freedom of George Eliot's *Daniel Deronda*', 67.
75. G. Eliot, *Daniel Deronda*, Book 8, p. 566.
76. For a discussion of 'anthropophagia' in the novel, see L. Toker (2004) 'Vocation and Sympathy in *Daniel Deronda*: The Self and the Larger Whole', *Victorian Literature and Culture* 32.2, 565–74, esp. at 569–70.
77. G. Eliot, *Daniel Deronda*, Book 8, p. 566, emphasis mine.
78. G. Eliot, *Daniel Deronda*, Vol. 4, Book 7, pp. 85–86, emphasis mine.
79. G. Eliot, *Daniel Deronda*, Vol. 3, Book 6, p. 253.
80. M. Arnold (1869) *Culture and Anarchy: An Essay in Political and Social Criticism* (London: Smith, Elder and Co.), p. ix.
81. G. Eliot, *Daniel Deronda*, Vol. 2, Book 4, p. 295.
82. B. Cheyette (1993) *Constructions of 'the Jew' in English Literature and Society: Racial Representations, 1875–1945* (Cambridge: Cambridge University Press), p. 5.

83. G. Eliot, 'The Modern Hep! Hep! Hep!', p. 221.
84. G. Eliot, *Daniel Deronda*, Book 8, p. 553.
85. G. Eliot, *Daniel Deronda*, Book 8, p. 553. According to Jane Irwin, in this passage, Eliot expresses her view that Judaism is a universalist religion and 'an original form of the Religion of Humanity'. See G. Eliot and J. Irwin, *George Eliot's Daniel Deronda Notebooks*, p. 132, note 3. The notebooks show that Eliot read several of Comte's works in preparation for the novel, including *Catichisme Positiviste, ou Sommaire Exposition de la Religion Universelle* (1852), *The Positive Philosophy of Auguste Comte* (1853), and the Positivist Calendar in *System of Positive Polity, or Treatise on Sociology, Instituting the Religion of Humanity* (1875). See G. Eliot and J. Irwin, *George Eliot's Daniel Deronda Notebooks*, p. 497.
86. G. Eliot, *Daniel Deronda*, Book 8, p. 545.
87. G. Eliot, 'The Modern Hep! Hep! Hep!', pp. 226–27.
88. G. Eliot, 'The Modern Hep! Hep! Hep!', p. 209. The question remains as to whether the settlement of 'the Jewish Question' that Eliot imagines, comes, as Monica O'Brien has suggested, at the expense of Anglo-Jewry's real historical and political possibilities, such that the quasi-religious, quasi-secular project that Mordecai and Daniel represent diverts 'the Jews' into a 'worldlessness', divests them of real worldly agency, and effectively exempts the British nation from confronting the Otherness of Jewishness and the problem that it poses for the peculiarity of British nationhood. Critiques such as O'Brien's see the proto-Zionism of the novel as a re-enactment of the racial exclusivity of nationalism at large, so that the Jewish people will necessarily impose the kind of national racial exclusion in their new homeland that has driven them from other nations. Further, O'Brien suggests that Eliot's focus on 'race' and willingness to usher 'the Jews' out of Britain indirectly contributed to anti-Semitism. Yet this critique, especially the charge that Eliot indirectly advanced anti-Semitism, fails to acknowledge the importance that Eliot places on national identity in general, for all peoples, and more importantly for my purposes here, the role that religion (as distinct from race, as I have shown) plays in producing that identity. See M. O'Brien (2007) 'The Politics of Blood and Soil: Hannah Arendt, George Eliot, and The Jewish Question in Modern Europe', *Comparative Literature Studies* 44.1/2, 97–117.
89. G. Eliot and G. S. Haight, *Letters*, Vol. 6, pp. 301–02.
90. See for example, T. M. Endelman (1999) *The Jews of Georgian England 1714–1830: Tradition and Change in a Liberal Society*, 2nd ed. (Ann Arbor: University of Michigan Press); A. Mufti (2007) *Enlightenment in the Colony: The Jewish Question and the Crisis of Postcolonial Culture* (Princeton, NJ: Princeton University Press); D. Boyarin, D. Itzkovitz, and A. Pellegrini (2003) *Queer Theory and the Jewish Question* (New York: Columbia University Press); S. Kadish (1992) *Bolsheviks and British Jews: The Anglo-Jewish Community, Britain, and the Russian Revolution* (London: F. Cass); C. Scheinberg (2009) *Women's Poetry and Religion in Victorian England: Jewish Identity and Christian Culture* (Cambridge: Cambridge University Press); B. Cheyette, *Constructions of 'the Jew' in English Literature and Society*; M. Scrivener (2011) *Jewish Representation in British Literature 1780–1840: After Shylock* (New York: Palgrave Macmillan); T. M. Endelman, *The Jews of Britain: 1656 to 2000*;

T. Kushner (1992) *The Jewish Heritage in British History: Englishness and Jewishness* (London: F. Cass); and M. Ragussis, *Figures of Conversion: The Jewish Question & English National Identity*.

91. One exception is an obscure and dated essay by S. Sharot (1971) 'Secularization, Judaism and Anglo-Jewry', in M. Hill (ed.) *A Sociological Yearbook of Religion in Britain: 4* (London: SCM Press), pp. 121–40. For a treatment of secularism and 'the Jewish Question' in Germany, see T. H. Weir (2014) 'The Specter of "Godless Jewry": Secularism and the "Jewish Question" in Late Nineteenth-Century Germany', *Central European History* 46, 815–49.
92. Bryan Cheyette uses the simple term 'semitism' to refer to the ambiguous, at once central and simultaneously marginal positioning of Anglo-Jews. See B. Cheyette, *Constructions of 'the Jew' in English Literature and Society*, p. 12.
93. M. Scrivener, *Jewish Representations in British Literature 1780–1840*, p. 6.
94. C. Scheinberg (1999) 'Re-Mapping Anglo-Jewish Literary History', *Victorian Literature and Culture* 27.1, pp. 115–24, at 118.
95. D. Englander (1988) 'Anglicized Not Anglican: Jews and Judaism in Victorian Britain', in G. Parsons (ed.) *Religion in Victorian Britain: I, Traditions* (Manchester: Manchester University Press in association with the Open University), pp. 235–73.
96. A. Benisch (1863) *Bishop Colenso's Objections: To the Historical Character of the Pentateuch and the Book of Joshua (contained in Part L) Critically Examined* (London: William Allan), p. vii.
97. S. Schechter (1911) *Studies in Judaism* (Philadelphia: The Jewish Publication Society of America), p. xii.
98. S. Schechter, *Studies in Judaism*, p. xiv.
99. S. Schechter, *Studies in Judaism*, p. xv.
100. Here I differ markedly from Englander, who suggests another analogy: 'In short, The "Historical School" was to Orthodox Jewry what *Essays and Reviews* and *Lux Mundi* were to Broad Church and High Church respectively'. This discounts Schechter's many acknowledgements of Catholicism and even the Oxford Movement, as well as the similar reliance on tradition as a living, evolving interpretation of Scripture on the part of the historical school and Catholicism. See D. Englander, 'Anglicized Not Anglican', p. 261.
101. S. Schechter, *Studies in Judaism*, p. xviii, emphasis in original.
102. D. Englander, 'Anglicized Not Anglican', p. 262.
103. G. Cantor and M. Swetlitz (eds), *Jewish Tradition and the Challenge of Darwinism*, p. 19.
104. G. Cantor, 'Anglo-Jewish Responses to Evolution', p. 26.
105. Jonathan Topham complicates this picture considerably in J. R. Topham (2004) 'Science, Natural Theology, and the Practice of Christian Piety in Early Nineteenth-Century Religious Magazines', in G. N. Cantor and S. Shuttleworth (2004) *Science Serialized: Representation of the Sciences in Nineteenth-Century Periodicals* (Cambridge, MA: MIT Press), pp. 37–66.
106. G. Cantor, 'Anglo-Jewish Responses to Evolution', p. 33.
107. G. Cantor, 'Anglo-Jewish Responses to Evolution', p. 34.
108. G. Cantor, 'Anglo-Jewish Responses to Evolution', p. 38.

Epilogue: Secularism as Modern Secularity

1. C. B. Upton (1910) 'Atheism and Anti-Theistic Theories', in J. Hastings (ed.) *Encyclopaedia of Religion and Ethics* (New York; Charles Scribner's Sons; Edinburgh: T & T Clark), Vol. 2, pp. 173–83.
2. E. S. Waterhouse (1921) 'Secularism', in J. Hastings (ed.) *Encyclopaedia of Religion and Ethics* (New York; Charles Scribner's Sons; Edinburgh: T & T Clark), Vol. 11, pp. 347–50, at 347–48.
3. E. S. Waterhouse, 'Secularism', p. 349.
4. E. S. Waterhouse, 'Secularism', p. 349.
5. E. S. Waterhouse, 'Secularism', p. 349.
6. G. J. Holyoake (1871) *The Principles of Secularism Illustrated* (London: Austin & Co.), p. 11, emphasis mine.

Index

142 Strand: A Radical Address in Victorian London, 90–91, 133, 169, 225

A

Abrams, Meyer H. (Mike), 19–20, 203, 205
Address to Men of Science, An, 12, 31, 36–39, 208–11
Agnostic Annual, 94, 120, 129, 131, 232, 234
agnosticism, 13, 18, 94, 109–10, 114–15, 117, 120, 122, 132–34, 141, 200, 228–29, 231–32, 234, 236
Anglo-Jewry, 189–90, 193, 196, 245–46
Anglo-Jews, 191, 193, 246
Arnold, Matthew, 52, 187–88, 244
Ashurst, William Henry, 89, 91, 118, 163, 204, 240
astronomy, 33, 36–37, 48, 59, 215, 218
atheism, 6, 9, 13–14, 18, 73–78, 81–82, 85, 92, 98, 100–101, 104, 109–10, 117, 125, 143, 145, 153, 159, 163, 165, 197, 222–25, 227, 231, 233
atheists, 34, 38, 75, 92, 98, 104, 126, 129, 159, 162, 164–66, 199, 204, 221, 225, 240–41
authority, scientific, 35, 216

B

BAAS (British Association for the Advancement of Science), 127–28, 233
Bacon, Francis, 40–41, 55
belief and unbelief, 8, 17, 43, 85, 108, 132, 166, 200, 204
Benisch, Abraham, 191–94, 246
Benn, Alfred William, 144–46, 149, 236–38
Besant, Annie, 95–97, 102, 226–27
Bible, 77–78, 85, 123, 142, 147, 149–53, 156, 160, 191–93, 204, 222, 236, 240–41
bigotry, 153–54
blasphemy, 30, 78–79, 84, 87, 112, 128, 204, 222–23, 231
Bonaparte, Felicia, 242–44
Bradlaugh, Charles, 9, 11, 13, 28, 73–74, 84–85, 94–95, 97–102, 105, 114–16, 124–25, 129, 197–98, 200, 208, 220, 223–27, 230–31, 233
Bradlaugh and Besant, 96–97, 102
Bradlaugh branch of Secularism, 11
Bradlaugh's atheism and neo-Malthusianism, 13, 110, 124
Bradlaugh's Secularism, 105, 197–98
Bradlaugh wing of Secularism, 123
Bristol, 35, 51, 76, 222
Britain, Christian, 193–94, 203
British Association for the Advancement of Science, *See* BSU
British science, 49, 54
British Secularism, 195, 233
British Secular Union. *See* BSU
Brougham, Henry, 33–34, 42, 50, 210, 212, 214–15
BSU (British Secular Union), 97, 121, 126, 226
Bynum, W. F., 145, 237–38

C

Cantor, Geoffrey, 194–95, 210, 213, 244, 246
Carlile, infidelity of, 74
Carlile, Richard, 9, 11–12, 16–19, 21, 23, 25, 27–43, 70, 73–76, 95–96, 101–2, 112, 200, 205, 208–12, 221, 227, 230
Carlile's deployment of scientific materialism, 35
Carlile's scientism, 17

Carlyle, Thomas, 9, 11, 17–18, 23, 25–26, 28, 32–33, 43, 79, 144, 151, 154, 206–8, 223
Carlyle and Carlile, 11–12, 16–19, 21, 23, 25, 27, 29, 31, 33, 35, 37, 39, 41, 43, 205
Catholicism, 14, 139–40, 158, 160, 192, 246
Chambers, Robert, 109, 113, 158
Chapman, John (publisher), 91, 223, 225, 236–37, 240–41
chemistry, 32–33, 36, 58, 182, 210, 217
Chilton, William, 112–13, 221, 224, 229–30
Christ, Jesus, 79, 140, 148–50, 152–53, 155, 162
Christianity, 2, 5, 10, 15, 18, 74, 78, 83, 125, 137–38, 143–44, 151–55, 160–63, 165, 170, 173–74, 180–81, 187, 190, 194, 202, 209, 226–27, 233, 241
  traditional, 10, 140, 150, 161–62, 236
Christians, 77, 149, 153, 164, 180, 190–91, 195
clothes, 22–25, 142, 164, 207
Clothes Philosophy, 22–23, 26, 206
Coleridge, Samuel Taylor, 20, 57, 209–10
colonialism, 63, 215
Comte, Auguste, 92, 118–19, 159, 225, 231, 241, 245
conduct, sexual, 103–4
Confidential Combination, 13, 90, 110, 123, 128, 133
configuration, religious-secular, 196
conjecture, 116–17
conscience, 137, 149, 157, 162
consciousness, 76, 123, 193
Consolations in Travel, 59, 61–62, 65–67, 69, 217–19
contraception, 95, 102, 124, 227
Conversations on Chemistry, 14, 58–59, 186, 217
conversion, 33, 83, 139, 141, 146, 164, 180, 200, 206, 241, 243, 246
convictions, religious, 35, 42, 138, 180, 194
Cooke, Bill, 131, 232, 234–35
crisis, religious, 135, 141
cultural authority, 35, 40, 69, 108
culture, national, 179

D
Darby, John Nelson, 148–49, 156, 238–39
Darwin, Charles, 2, 14, 97, 120, 123–24, 129, 138, 182, 226–27, 229, 232–33
Darwinian science, 193, 195
Darwinism, 9–10, 97, 123, 130, 191, 193–95, 244, 246
  social, 34, 181–82, 191
Darwinism and scientific naturalism, 195, 242
Davy, Humphry Sir, 33, 35–36, 39, 58–59, 61–62, 64–66, 69, 209–11, 215, 217–19
Dawson, Gowan, 97–98, 115, 123–24, 211, 226–29, 231–33, 235
Deronda, Daniel, 15, 172–74, 181, 184, 186–88, 189–90, 195–96, 241–45
desecularization, 19, 21, 27, 202, 206, 212
Desmond, Adrian, 60, 109, 131, 133, 233
determinism, 181, 183–84
development
  progressive, 61, 66–67, 219
  successive, 67, 219
dialogues, 62–65
discourse, religious, 5, 9, 14, 20, 105, 137–38
disenchantment, 21, 26–27, 140, 146, 207
disproof, 83, 92
diversity, increasing religious, 20, 43
divine Unity, 188–89
doctrine
  atheistical, 64–65
  religious, 33, 118, 165, 236

E
Eclipse of Faith, The, 155–56
Edinburgh Review, The, 50–51, 214
education, scientific, 39–40, 210
elements, secular and religious, 6, 24, 199

Eliot, George, 9–10, 13–15, 20, 91, 94, 160, 169–79, 181–85, 187, 189–91, 193, 195, 204, 206, 225–26, 235, 239–45
Eliot's fiction, 15, 169–70
Eliot's persona, 179
Eliot's portrayal of Judaic religiosity, 187
Eliot's post-secular narratives, 170
Eliot's secular outlook, 240
empire, 51–52, 62–63, 176
enchantment, 21, 26–27, 146
Encyclopaedia of Religion and Ethics, 197, 247
Englander, David, 191–93, 246
English Secularism: A Confession of Belief, 115, 220, 223–24, 231
Enlightenment rationality, 8–9, 16, 19, 28, 105
Epstein, James A., 39, 208–9, 211
Essay in Aid of a Grammar of Assent, An, 138, 237
eternity, 4, 23–24, 118, 150, 170, 207
Ethics, 184
Evangelical Disenchantment: Nine Portraits of Faith and Doubt, 140, 235, 237–40
evangelicalism, 18, 20, 140–41, 144, 146, 153–54, 237, 239
evangelical Protestantism, 138–41
Evans, Marian, 91, 93, 160, 225
Every Woman's Book, 101–2, 227
evidence, presumptive, 82
evolution, 47, 49, 60, 156, 181–82, 193
evolutionary science, 124, 140, 154, 194–96
evolutionary theory, 61, 69, 108, 110, 114, 124, 157, 193

F
faith, crisis of, 2–3, 15, 18–21, 71, 146, 155, 190, 193, 200, 203, 205, 212
Family Library, 56–57, 59, 217
fellowship, 147, 178–79, 190
Foote, George William, 115, 121, 126, 232
fossil series, 47, 60–61, 65
freethinking atheism, 117
freethought, 71, 73–75, 80–85, 87, 92, 97–99, 103, 105–6, 114–16, 118–19, 124, 127, 129, 133, 137, 164, 167, 205, 208–9, 211–12, 220–22, 224, 229–30, 234, 236
freethought movement, 6, 43, 74–75, 85, 95, 97–98, 111, 116, 128, 153, 163
Fruits of Philosophy, 95–97, 102, 124, 226–27

G
genius, 39–40, 62–63, 189
geology, 9–10, 12, 44, 47–49, 58–61, 63, 65, 151, 213–14, 217–18
  history of, 47–48, 213
George Holyoake and Modern Atheism, 164, 240
George Holyoake's Secularism, 11, 169
God, 4, 14, 22, 29, 59, 63, 82, 112, 137–39, 146, 148–51, 153, 160–62, 171, 173, 186, 195, 198, 202–4, 215, 219, 222, 240

H
Habermas, Jürgen, 13, 202, 205, 228
heaven, 23, 33, 83, 161
Hempton, David, 140, 162–63, 235, 237–40
Henriques, Alfred Gutteres, 182, 195, 244
Hieroglyphical Truth, 207
Higher Criticism, 154, 190–93, 195, 242
historic memories, 188
historiography, 18, 108, 111, 120, 132–33, 228
history is not religion, 152
history of freethought, 167, 220–22
Holyoake, peculiar religiosity influence, 165
Holyoake, Austin, 96
Holyoake, George Jacob, 3, 5, 8–9, 11–14, 16, 71–74, 77–94, 96–101, 103–6, 108–12, 114–19, 121, 123–30, 132–33, 136–38, 142–43, 150, 163–67, 169, 197–99, 220–27, 229–35, 238, 240–41, 247
Holyoake & Co (publisher), 230–31

Holyoake and Bradlaugh camps, 104, 115, 124
Holyoake and company, 6, 13, 80, 86, 106, 204
Holyoake and Secularism, 71, 73, 75, 77, 79, 81, 83, 85, 87, 89, 91, 93, 95, 97, 99, 101, 103, 105, 219
Holyoake branch of Secularism, 11
Holyoake brand of Secularism, 232
Holyoake camp, 103, 124, 126
Holyoake camp of Secularism, 123
Holyoake's brand of Secularism, 11, 13, 74, 123
Holyoake's conception of Secularism, 104
Holyoake's contribution to freethought and scientific naturalism, 129
Holyoake Secularist, 126
Holyoake's History, 127
Holyoake's position, 99, 102–3
Holyoake's *Reasoner*, 98
mocked, 98
Holyoake's reasoning, 100
Holyoake's review, 163–64
Holyoake's role, 93, 133
Holyoake's Secularism, 9, 12–15, 70, 72–74, 98, 103–6, 109–10, 116, 123–25, 132–33, 198, 200, 241
preferred by scientific naturalists, 111
Holyoake's Secularism and scientific naturalism, 123
human nature, 83, 118–19, 165
Hunt, Thornton, 13, 89, 165, 169
Huxley, Leonard, 131, 234
Huxley, Thomas H., 13, 35, 52, 82, 91, 93, 108–10, 114, 120, 122–25, 127–33, 129, 195, 225–26, 231–35

I
ideology, 31, 36, 43, 55, 134, 211, 220, 229, 233–34
religious, 32, 74, 81
infidelity, 12, 14, 28, 73–74, 76, 80, 110, 115, 124, 133, 140, 143, 155, 164, 221, 235, 239
infidels, 38, 74–76, 83, 129, 165, 232, 234–35, 240
institutions, 6, 31, 40, 51, 55, 109, 215
intelligence, 88, 118, 122, 127
Investigator, 98–99, 125, 227
Ireland, 53, 147–48, 232

J
Jewish, 173–74, 179, 182–83, 186–90, 245
Jewish Chronicle, 182, 191–92, 194
Jewish identity, 175–76, 183–85, 245
Jewishness, 183, 242, 245–46
Jewish Question, 245–46
Judaic, 173
Judaic faith, 180, 183
Judaic religiosity, 172, 187
Judaism, 15, 153, 172–74, 176, 180–81, 183, 187, 189–96, 201, 245–46

K
Klancher, Jon P., 49, 209, 214
Knight, Charles (publisher), 56–57, 109
knowledge
diffusion of, 42, 73, 80
natural, 51–52, 54–55, 163, 209
plebeian, 40–41
secularize, 42
knowledge industry, 40, 42, 49, 56, 69
knowledge production, 12, 50, 52, 69
Knowlton, Charles, 95–96, 124, 226–27
Knowlton affair, 95–97, 103, 115, 121, 125–26, 226

L
labor, 39–40, 127, 157, 163
Larsen, Timothy, 18–19, 203, 205, 236–37
laws, existing, 64–67
Leader, 90–91, 93–94, 98–99, 123, 165, 169, 236
Lewes, George Henry, 13, 89–94, 103, 129, 165, 169–70, 225, 241–42
Library of Entertaining Knowledge, 56
Library of Useful Knowledge, 56–57
life
animal, 64, 67
organic, 61, 66–67, 123
religious, 4, 198
Lightman, Bernard V., 115, 120, 130, 210, 216, 219–20, 226, 228–35

Literary Secularism, 183, 244
Logic of Scientific Discovery, The, 118, 231
London Central Secularist Society, 99
London Investigator, The, 227
London University, 50–51
London University, The, 214
Longman, 59, 203, 209, 212, 230
Lyell, Charles, 9, 12, 47–56, 58–61, 63–69, 213–19
Lyell-Murray knowledge project, 47, 49
Lyell's geological science, 11
Lyell's history, 214
Lyell's *Principles*, 50, 58–61, 68–69, 151, 213, 218
Lyell's project, 49
Lyell's science, 50, 69
Lyell's secularism, 69
Lyell's uniformity, 69–70

M
Manwaring, Kathleen, 146–47, 237–38
marriage, 35, 103, 162, 174–76
Marx, Karl, 76–77, 222
material conditions, 29, 75, 79–80, 82, 165
materialism, 17, 29–30, 34, 36–37, 60, 63, 75, 80–82, 112, 116, 125, 127, 218, 230–31
materialists, 31, 34, 38, 60, 63, 79, 111–14, 129
matter, absence of, 77
mechanics, 39–41, 86, 212
Mechanic's Magazine, 38, 40–42, 55, 212
Messiahship, 152–53
mid-century Secularism, 3, 11, 13, 16, 74, 104, 106, 132, 200
Middlemarch, 170, 176, 241–43
Miller, J. Hillis, 25, 181, 207
miracles, 27, 151–52, 236
modern secularity, 3–4, 7–8, 11, 13–15, 19, 71–74, 197, 200, 247
Moral Physiology, 102, 227
moral system, 74, 81, 83, 165, 177, 199
More Nineteenth-Century Studies: A Group of Honest Doubters, 135, 140, 203, 235–38
Murray, John, 53, 55–61, 67–69, 216–17

N
Nash, David, 202–5, 208, 212, 228
National Reformer, 11, 96, 99–101, 125, 224
National Secular Society. *See* NSS
Natural History, 14, 49, 51, 59, 109, 113–14, 138, 156, 215, 217, 221, 229, 236, 239
naturalism, 31, 61, 67, 81–82, 109–10, 121, 224
Naturalist, 52, 54, 121, 232
natural philosophy, 37, 49, 51–52, 59–60, 211
  study of, 37, 54, 59, 217
natural sciences, 51–52, 54, 121, 192, 214
  contemporary, 194
natural supernaturalism, 9, 17, 19, 21–23, 26–28, 203, 205–7
natural theology, 22, 31, 35, 37, 49, 52, 54, 59–60, 83, 194, 205, 210–11, 215, 218, 246
NCA, 225, 230, 232–35
Negative Atheism, 101
Neo-Malthusian doctrine, 102–3
new knowledge industry, 42, 49, 56–58, 61, 69
Newman, 9, 14, 89–90, 129, 135–41, 143–49, 151–65, 167, 225, 235–41
Newman, Charles Robert, 14, 136, 139, 142–43, 235, 237
Newman, Charles Robert life history, 142
Newman, Francis William, 9, 13–14, 20, 89–91, 93, 99, 129–30, 136–38, 140–50, 152, 155–56, 163–64, 166–67, 170, 235–37, 240
Newman, Francis Willliam naturalization, 14, 138
Newman, John Henry, 14, 129, 135–36, 138–43, 147, 152, 155–56, 166, 237–38
Newman Brothers, 14, 135–36, 140, 145, 235, 237
Newman's crisis, 145–46
Newman's refutations of atheism, 159
Newman's secularism, 145
Newman's theism in Holyoake's review, 164

New Reformation, 90–91, 137, 165, 169
New Testament, 149–53, 191
New York Tribune, 127, 233
nineteenth-century geology, 47–48
nineteenth-century religiosity, 136
nineteenth-century science, 120
nineteenth-century Secularism, 45, 72
nineteenth-century secularity, 137
novel, secular-religious, 15
NSS (National Secular Society), 11, 84, 97, 100, 114–15, 121, 126, 130, 226, 233

O
obscenity, 95–96, 115
Old Testament, 151–52, 191
oppression, 31, 40, 178
optative condition, 17, 43, 166–67, 200
Oracle of Reason, 76–81, 84, 111–13, 116, 221–23, 229–30
Owen, Robert, 74, 76, 102, 118, 221
Owen, Robert Dale, 102, 227
Owenism, 11, 73, 76, 80, 89, 109, 221

P
Paul (apostle), 149, 153, 162
Pentateuch, 151, 192, 246
Phases of Faith, 14, 91, 135, 137–38, 141, 144, 146–47, 153–55, 165, 236–39
philosophy, materialist, 26–27
phrenology, 109
physical sciences, 55, 59, 113, 121, 151
pluralism, secular-religious, 200
pluralities, 7, 43
Positive Philosophy, 92, 245
positive principles, 100, 198
post-secular, 2, 9, 11, 13, 105, 171, 202, 205
post-secular condition, 1, 4, 7, 15, 171, 200, 202
post-secularism, 4, 6–7, 9, 15, 171–73, 202, 204
post-secularist, 15, 169
Principles, 48–49, 54–59, 61, 63, 65, 67, 115, 214, 218–19
Principles of Geology, 10, 12, 45, 47, 49, 51, 53, 55, 57, 59, 61, 63, 65, 67–69, 212–19
Principles of Secularism Briefly Explained, 115, 230
Principles of Secularism Illustrated, The, 199, 220, 230, 247
project, secular-religious, 169, 241
proofs
  scientific, 139
  syllogistic, 161
Protestant, 138–40, 160, 239
Protestantism, 136, 139–40, 152, 160
public sphere, 5–6, 12, 30, 34–35, 37, 45, 47, 74, 86, 105, 108, 202, 210

Q
Qualls, Barry V., 206–7, 212
Quarterly Review, 50–53, 55–56, 58, 60, 68–69, 209, 214, 216, 228

R
Radical Expression, 29, 208–9, 211
radicals, 34, 71, 75, 123–24, 203, 219–20, 226–27, 234
Radicals, Secularists, and Republicans: Popular Freethought in Britain, 1866–1915, 71, 219–20, 226–27, 234
rationalism, 11, 14, 17–18, 136, 144–45, 191, 235–36
Rationalist Press Association. *See* RPA
Reasoner, 5, 74, 76, 83–89, 91–93, 96, 99, 114–18, 123, 136, 142–43, 163–64, 204, 224–27, 230–31, 236, 240–41
Rectenwald, Michael D., 107, 168, 204–5, 219, 221, 229, 231, 241
reductionism, 32–33
re-enchantment, 21, 27–28, 146, 200, 206, 208
reformist article, 51–53
registers, visceral, 173–74
relationship, science-religion, 35
religion
  absence of, 5, 106
  evangelical, 141, 237
  new, 91, 165
  persistence of, 6–8, 171
  positive, 170, 241
  rational, 137, 221
  science versus, 210, 216

religionists, 165–66, 216
religion of the heart, 137, 236
religiosity, 2–3, 5, 8, 12, 74, 105, 138, 146, 154, 158, 162, 171, 174, 181, 184, 186–87, 191, 200, 203–4, 212, 223
  new, 14, 164
  new kind of, 147, 157, 165
religious belief, 7–8, 10–12, 16, 18–19, 28, 30, 85, 101, 105, 138, 170–71, 173, 194
religious believers, 1, 9, 105, 164, 170
religious bigotry, 153
religious commitments, 146, 173
religious elements, 6, 24, 199
religious experience, 7, 157–58, 162
religious faith, 10, 17
  existing, 20
religious history, 142
  reshaping of, 9
religious ideas, 77, 236
religious interlocutors, 13, 110, 124
religious orthodoxy, 181, 210
religious persons, 6, 154
religious school, 156, 236
religious sects, 43, 105
religious sentiment, 19, 129, 169
  valued, 130, 169
religious terms, 18, 28
religious turn, 2, 15, 172, 203
religious vestures, 24
religious worship, 93, 160, 226
Republicans, 10, 28–29, 71, 203, 208, 211, 219–21, 226–27, 234
researches, 33, 35–36, 50, 106, 195
resistance, 75, 172, 183–84, 186
revolution, 32, 49, 60, 77, 203, 205, 214
reward, 23, 83, 171
rhetoric, 35, 74–76, 78, 80, 84, 86, 180, 206, 208
Robbins, 146, 202, 209–10, 235, 237
Romanism, 136, 140, 152
Romanticism, 2, 19–20, 26, 119, 203, 208, 215
Romanticism/Secularization/Secularism, 205
Romantic secularization, 17, 19
Royal College of Surgeons, 29, 32, 51, 112, 209
Royle, Edward, 71–72, 87, 90, 98, 105, 130, 203, 211, 219–30, 233–34, 236
RPA (Rationalist Press Association), 129–32, 232, 234
Ryall, Maltus Questell, 81, 222–23

S

salvation, 83, 152–53, 170
Sartor Resartus, 11, 17, 19–22, 24–28, 154, 206–8
  natural supernaturalism of, 19
Scenes of Clerical Life, 168–69
Schechter, Solomon, 192–93, 246
schools, historical, 192, 246
science
  historians of, 35, 48, 132–33, 194, 211, 213, 215
  man of, 35, 122, 225, 232
  materialist, 10, 45
  men of, 29, 37
  modern, 11, 45, 120, 154, 195
  paradigm-shifting, 11, 193
  political, 118
  radical, 40, 60, 70
  revolutionary, 10, 40
  secularist, 31
  secularize, 134
  secularizing, 228
  theistic, 120–21, 228
science and secularism, 37
science versus religion controversy, 132
science versus religion disputes, 52
Science Wars, 30
Science Wars, 209
Scientific Institutions, 51–52, 55, 214–15
scientific knowledge, 28, 30–31, 39, 45, 58, 110, 212, 217
  promoted, 42
scientific knowledge production, 12, 45, 52
scientific knowledge project, 12
scientific materialism, 16, 23, 28, 30, 34–37, 211
scientific meliorism, 94, 226
scientific methods, 48, 118

scientific naturalism, 9, 13, 35, 69, 82, 108–11, 117, 119–24, 129–30, 132–34, 171, 194–95, 200, 210, 216, 219, 228–29, 232, 242
  acceptance of, 14, 110
  creed of, 13
  early stage of, 14, 110
  emergence of, 9, 14, 109, 111, 113, 115, 117, 119, 121, 123, 125, 127, 129, 131, 133, 228
  promotion of, 109, 124
scientific naturalists, 13, 49, 52, 97, 109–11, 115, 120, 122–28, 130, 132–34, 194–95, 216, 228
  prominent group of, 13, 110
scientific practice, 120, 134
scientific revolution, 10, 34, 49, 51, 215
scientific worldview, 120–21
scientists, 29, 32, 118
Scriptures, 139–40, 147, 149, 151–52, 192, 246
SDUK (Society for the Diffusion of Useful Knowledge), 12, 56–57
Secord, James A., 111, 113, 213–19, 221–22, 229–30, 239
secular affairs, 4
Secular Age, A, 7, 45, 72, 203–5, 212, 220, 242–44
secular conception, 198
secular conversions, 20, 43
secular framework, 15, 169
secular improvement, 14, 100, 165, 170
secular interventions, 3, 9
Secularism, 1–6, 8–9, 11–16, 18, 20, 71–75, 83, 85, 87–95, 97–101, 103–6, 108–11, 113–19, 121–25, 129–34, 137–38, 163–67, 169–71, 197–201, 203–6, 219–21, 223–29, 231–34, 239–42, 246–47
  accounts of, 72
  definition of, 98, 125, 227
  development of, 9, 82, 89, 164, 170
  differentiated, 98, 220
  hard, 12, 200
  mid-nineteenth-century, 9
  modern, 71, 73, 104, 116
  movement of, 12, 72, 95
  organized, 72
  philosophical, 171, 241–42
  principles of, 111, 115, 117, 229
  scientistic, 205
  term, 92
  understanding of, 73, 116
  version of, 8, 115
  western historical, 8
Secularism and freethinking atheism, 117
Secularism and scientific naturalism, 123, 132
secularism and secularization, 2, 190
Secularism by Holyoake and company, 106
secularism/religion, 6
Secularism/secularism, 9, 14
Secularism's history, 130
Secularist camps, 97–99, 124
Secularist movement, 74, 95, 97, 191, 200
Secularists, 69, 74, 91, 97–98, 100–101, 103, 105, 109–10, 115, 117–18, 121–22, 124–26, 128, 163, 165–66, 170, 187, 198, 203, 224, 230, 232–33, 240, 242
secularity, 1, 3–4, 7–9, 11, 13–17, 19, 21, 43–44, 104, 106, 108, 134–39, 141, 143, 145, 147, 149, 151, 153, 155, 161–63, 167, 172–73, 194–96, 200–203
  condition of, 14, 21, 108, 171, 200
  emergent, 3, 34
  new, 12, 28
  optative condition of, 43, 136, 166
  overarching, 73, 85
  sense of, 7, 13, 105
  understanding of, 106, 200
secularization, 1–4, 6–7, 9–11, 15, 17–21, 28, 45, 71–72, 105–6, 108, 132, 134, 144, 160, 162, 172–73, 181, 187, 190–91, 193–96, 200–203, 205, 212, 219–20, 222–23
secularization process, 10, 160
secularization theorist, 202, 212
secularization thesis, 1–3, 6–7, 19–20, 202
secularizing, 10, 35, 111, 134, 176, 190, 193, 196
secularizing tendencies, 36–37

secular knowledge, 101
secular life, 165, 198
secular matter, 24, 198
secular mode, 19, 28
secular modernity, 3, 18
secular movement, 5, 39, 84
secular outlook, 69, 202
secular press, 123, 133
Secular principles, 101, 117–18
Secular Review, 97, 121, 126, 129, 232–33
Secular Review and Secularist, 121, 230, 232–33
secular science, 35
Secular societies, 7, 43, 203
secular sublime, 170, 175, 240
secular system, 104, 118
sentiments, moral, 137, 190
separateness, 179, 187–89, 242
sexual policy, 97–98, 103–4, 124–25
Short History of Secularism, A, 72, 220, 242
Singh, Amardeep, 183, 244
Smith, Graeme, 72, 220, 242
social science, 80
Society for the Diffusion of Useful Knowledge. *See* SDUK
Soul: its Sorrows and Aspirations: The Natural History of the Soul, as the True Basis of Theology, The, 14, 90, 91, 135, 137–38, 146–47, 156–57, 160, 162–66, 236, 239–41
Southwell, Charles, 73, 76–78, 80–82, 84–85, 98–99, 111–12, 222–24, 227, 229
species, 10, 32, 59, 61, 64, 66–68, 109, 112, 157
  extinct, 65–66
species transmutation, 60, 64–67, 109
Spencer, Herbert, 13, 89–91, 93, 108, 110, 120, 123–24, 127, 130, 132–33, 165, 169–70, 181, 225, 233, 235, 237, 240–41, 243
spheres, religious, 21, 104, 106
Spinoza, Baruch, 82, 184–85, 244
spiritualism, 18, 121, 239
standard secularization thesis, 1–3, 7, 16–19, 26–28, 45, 73, 104–6, 145, 196, 208, 228
State of the Universities, 53, 214–16
state religion, official, 29, 104
strata, 63–66
superstition, 28, 31–32, 79–80, 135, 154, 182, 208–9, 236
Symbols, 25–26
system, religious, 101, 104

T
Taylor, Charles, 7–8, 17, 21, 45, 125, 203–6, 212, 220, 225, 233, 242
Taylor, G. H.
Teufelsdröckh, Diogenes, 22, 24–27, 206
theism, 12, 14, 18, 74, 78, 82, 85, 114, 117, 119, 164, 197, 199
theists, 13, 85, 104–5, 120, 122, 125, 137, 159, 164, 199, 216
theory of regular gradation, 111–13, 230
T. H. Huxley Papers (HP), 233–35
this life, 4–5, 81, 118, 134
The Three Newmans, 167, 240
transcendence, 15, 27, 172–73, 187–89
  cultural, 188–90
trial, 78, 93, 95–96, 115, 128, 222
Trinity, 149
Turner, Frank M., 20, 43, 45, 108–9, 119–20, 141–42, 203, 206–7, 210, 213, 216, 219, 228, 232, 237–38
Tyndall, John, 13, 110, 121, 123–24, 126–29, 131–33, 195, 210, 216, 219, 228, 232–35

U
uniformity, 47–48, 61, 64–70, 213

V
Varieties of Religious Experience, The, 157
Vestiges of the Natural History of Creation, 59, 109, 113–14, 217, 221, 229–30, 239
Victorian science, 108, 120

W
Watts, 58, 96–97, 100, 121, 129–32, 230–32, 234
Watts, Charles, 96–97, 115, 120–21

Watts, Charles Albert, 94, 120, 129, 232
Watts, Charles Secularist, 94, 129
Westminster, 93–94, 133, 226
Willey, Basil, 140, 146, 203, 235–38
Wilson, Leonard G., 65, 213–19
word history, 111, 114–15, 119
working-class freethought movements, 93–94
worship, 160, 170

Z
Zetetic movement, 37–38, 40, 42, 211
Zetetics, 38–40
Zetetic societies, 12, 38

Made in the USA
Coppell, TX
09 July 2020